Aims and Scope

The series "Springer Theses" brings together a selection of the very best Ph.D. theses from around the world and across the physical sciences. Nominated and endorsed by two recognized specialists, each published volume has been selected for its scientific excellence and the high impact of its contents for the pertinent field of research. For greater accessibility to non-specialists, the published versions include an extended introduction, as well as a foreword by the student's supervisor explaining the special relevance of the work for the field. As a whole, the series will provide a valuable resource both for newcomers to the research fields described, and for other scientists seeking detailed background information on special questions. Finally, it provides an accredited documentation of the valuable contributions made by today's younger generation of scientists.

Theses are accepted into the series by invited nomination only and must fulfill all of the following criteria

- They must be written in good English.
- The topic should fall within the confines of Chemistry, Physics, Earth Sciences, Engineering and related interdisciplinary fields such as Materials, Nanoscience, Chemical Engineering, Complex Systems and Biophysics.
- The work reported in the thesis must represent a significant scientific advance.
- If the thesis includes previously published material, permission to reproduce this must be gained from the respective copyright holder.
- They must have been examined and passed during the 12 months prior to nomination.
- Each thesis should include a foreword by the supervisor outlining the significance of its content.
- The theses should have a clearly defined structure including an introduction accessible to scientists not expert in that particular field.

Michael Steger

Transition-Metal Defects in Silicon

New Insights from Photoluminescence
Studies of Highly Enriched ^{28}Si

Doctoral Thesis accepted by
Simon Fraser University, Burnaby, BC, Canada

 Springer

Author
Dr. Michael Steger
Department of Physics
Simon Fraser University
Burnaby, BC
Canada

Supervisor
Dr. Michael L. W. Thewalt
Department of Physics
Simon Fraser University
Burnaby, BC
Canada

ISSN 2190-5053   ISSN 2190-5061   (electronic)
ISBN 978-3-642-43808-0   ISBN 978-3-642-35079-5 (eBook)
DOI 10.1007/978-3-642-35079-5
Springer Heidelberg New York Dordrecht London

for dad

Supervisor's Foreword

Devices made from the semiconductor silicon power our ubiquitous computing, communication, information, power, and entertainment technologies. Enormous expenditures have been made in silicon materials and nanofabrication technology, and in the purification and perfection of silicon crystals. It would be surprising if such a thoroughly studied and widely exploited material could reveal novel and unexpected properties at this late date, but that is exactly what this thesis is about.

Silicon, along with almost all other semiconductors, is found in nature as a mixture of stable isotopes, in this case ^{28}Si, ^{29}Si and ^{30}Si. Changing this isotopic composition can have subtle effects on some of the properties of semiconductors, as has been known for some time from studies of isotopically enriched germanium and diamond, but samples of isotopically enriched silicon did not become available until quite recently, as a result of the Avogadro Project, which seeks to redefine the kilogram based on the properties of highly pure and highly enriched ^{28}Si.

Along with the subtle effects previously observed in other enriched semiconductors, the ^{28}Si revealed something completely new: the linewidths of many different optical transitions, spanning the near-gap region to the far-infrared, were found to be much sharper in ^{28}Si than in natural Si. This was an exciting result for semiconductor spectroscopy, since the linewidths even in natural Si were already much sharper than in other semiconductors, due to the purity and perfection of Si. The origins of the inhomogeneous isotope broadening which dominates in natural Si were quickly understood, but until their experimental discovery it had been thought that such effects would be too small to be significant.

Michael Steger's thesis describes in exquisite detail one important aspect of these incredibly narrow spectral lines which can be observed in ^{28}Si. For several decades researchers had been studying deep luminescence transitions (that is, transitions well below the $\sim$1.15 eV band gap energy) and had associated some of these deep luminescence centers with various transition metal impurities. This had practical implications, since transition metals are typically very deleterious for device performance, so detecting small quantities of them in Si was important. Over the years these identifications became well established in the literature, the prime example being the Cu-pair center. This identification was so widely accepted that the Cu-pair center was even used as a test bed for ab initio defect calculations comparing the calculated properties of a defect with its known properties.

Michael discovered finestructure in the luminescence of these centers, which he named isotopic fingerprints. These fingerprints unambiguously identify the type and number of atoms in a complex defect. Beginning with the 'Cu-pair' center, which was shown to actually contain four Cu atoms, it was found that not a single one of the previously studied deep defect centers is actually what everyone up until then thought it to be. This led to consternation in the deep defect community, but the evidence is so convincing that no one has tried to argue with the new identification made using isotopic fingerprints in ^{28}Si. This thesis provides a comprehensive listing of the known properties of these newly identified defects, which comprise sets containing either four or five atoms chosen from among Cu, Li, Au, Ag, and Pt. This information will be invaluable in the much-needed modeling of these centers, in order to understand their formation and properties, and explain why the number of constituents is always four or five. There is little doubt that other, less well-studied deep luminescence centers in Si will reveal new surprises when their isotopic fingerprints are studied in ^{28}Si.

Burnaby, October 2012   Dr. Michael L. W. Thewalt

Acknowledgments

First and foremost, I would like to thank my Senior Supervisor, Dr. Michael L. W. Thewalt, for accepting me into his research group. His vast knowledge, his intuitive experimental skills, and his excitement for science provided invaluable support and guidance for this work. Furthermore, I acknowledge my colleagues in the lab, Albion Yang, Kamyar Saeedi, and Takeharu Sekiguchi.

This work was supported by the Natural Sciences and Engineering Research Council of Canada (NSERC). The author would like to thank M. O. Henry and K. Johnston for Au implantations, E. Alves and U. Wahl for Pt implantations, and S. K. Estreicher, J. Weber, E. E. Haller, M. O. Henry, and G. Davies for helpful discussion.

This project would not have been possible without the sample material provided by the Avogadro Project.

For generous financial support, I would like to acknowledge Dr. Michael L. W. Thewalt and the Department of Physics, Simon Fraser University.

And last, but not least, I am especially indebted to my father, Maximilian, and to my wife, Garima.

Contents

Chapter 1
Introduction and Background

1.1 Introduction

1.1.1 Silicon

The silicon (Si) crystal has the diamond structure, in which each Si atom is placed at the centre of a tetrahedron consisting of its four nearest neighbours. In doing this, Si creates a face-centered cubic lattice (fcc) with its two basis atoms located at the origin $(0,0,0)$ and at one quarter length along the diagonal $[111]$ direction at $(\frac{a}{4}, \frac{a}{4}, \frac{a}{4})$, with $a = 5.43\,\text{Å}$ [1] being the lattice constant of Si. The arrangement of Si atoms in this diamond lattice is shown in Fig. 1.1.

The study of impurities in Si has always been of strong interest, as the electrical properties can be varied greatly by introducing controlled amounts of certain different elements into the Si crystal. These impurities can take either of two places in the Si crystal: They can be situated at a substitutional site, that is, they replace one of the Si atoms in the crystal, or, the impurity can be located interstitially, that is, in the space in between the regular fcc-crystall lattice sites. Depending on the electronic levels created by the impurity in the bandgap of the semiconducting Si crystal, the impurity can be characterised as shallow (e.g. phosphorous or boron, where the electronic level is close to the conduction or valence band edge, respectively) or deep, like the impurities that will be discussed here. As we will see, it is often difficult to determine the exact location or arrangement (the crystallographic symmetry) taken by the impurity in the crystal. Because the inclusion of an impurity into the crystal lattice changes the crystal's perfection, they are often referred to as defects. Many, but not all those defects have electronically excited states, whose decay can be observed by the emitted photoluminescence (PL). The energy of the emitted photons is characteristic for many of the defect's properties.

M. Steger, *Transition-Metal Defects in Silicon*, Springer Theses,
DOI: 10.1007/978-3-642-35079-5_1, © Springer-Verlag Berlin Heidelberg 2013

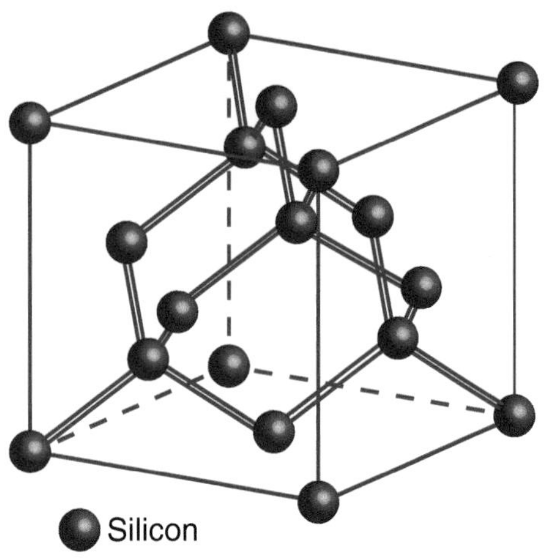

Fig. 1.1 A unit cell of the face-centered cubic lattice with silicon basis atoms at (0,0,0) and at $(\frac{a}{4}, \frac{a}{4}, \frac{a}{4})$, with $a = 5.43\,\text{Å}$ [1] being the lattice constant of silicon

1.1.2 Transition Metals in Silicon

Deep photoluminescence (PL) centres associated with transition metals (TM) in Si have been studied for several decades. There has been interest in these centres as markers for the detection of various transition metal contaminants, which have deleterious effects on semiconductor device performance, even at very low concentrations [2–4]. The strong PL of TM impurities in Si even at small concentrations (high intensity and efficiency) [5], the rapid diffusion of some of the impurities [6] and the possibility to use these efficient isoelectronic bound exciton (IBE) PL centres as a Si-based light source [7] have led to a significant research interest over time. Several different characterization methods have been used on these centres:

- PL under strain or in a magnetic field to determine their crystallographic symmetry [8, 9]. In these experiments, external strain is applied to the crystal in different crystallographic directions. The strain lifts the degeneracies of different electronic levels and a splitting of the PL spectrum is usually observed;
- PL excitation experiments to study the centre's excited states [10]. Light is used to populate highly excited electronic states;
- Controlled diffusion with specific, pure metals to produce certain centres, which did not eliminate the often unrecognized problem of unintentional contaminants [11];
- Controlled diffusion of specific isotopes into silicon to observe a small isotope shift [9]. The PL energy depends on the isotopic mass of the impurity;
- The relationship between impurity content and the observed PL intensity [9, 12]. It is expected that a higher concentration of the same defect sites generates a stronger signal, but it is difficult to determine the actual impurity content of a sample;
- The use of radioisotopes to determine the involvement of specific elements by observing the PL intensity with respect to the decay time [13]. This method is

complicated by the necessary use of radioisotopes, but the decay and growth of different defect sites can be observed;

- The connection between TM PL centres and corresponding deep level transient spectroscopy (DLTS) centres [14]. While no direct connection between these two experiments can be made it is sometimes possible to find corresponding features in the spectra.

However, while much has been learned about these centres, none of these methods could give a definitive answer regarding the number of atoms of a given element contained in a centre, and the actual composition often remained under debate.

1.1.3 Isotopically Enriched Silicon

The resolution of many PL features is inherently limited by the observed spectroscopic width of the PL line in the spectrum. Recently, optical studies of isotopically enriched ^{28}Si have given rise to a number of unforeseen results due to the significantly reduced spectroscopic linewidths, which is a result of the elimination of inhomogeneous isotope broadening [15, 16] inherent to ^{nat}Si. The inhomogeneous isotope broadening is often the determining factor in the observed linewidth. These results span the fields of shallow [16–21] and deep [22, 23] impurity infrared absorption spectroscopy, shallow bound exciton (BE) PL [15, 19, 24–26], photoluminescence excitation (PLE) spectroscopy [19, 27–31], and, as discussed in this work, deep centre PL spectroscopy [32–37]. The deep PL centres are often found to be isoelectronic bound exciton (IBE) centres with a relatively large PL efficiency and a long lifetime when compared to shallow donor and acceptor BE in Si. The reduced spectroscopic linewidths have resulted in a new and powerful method of characterizing the constituents of these deep IBE centres in Si, as is demonstrated in Chap. 4. It is now possible to observe so-called isotopic fingerprints. Such a fingerprint is a complicated pattern in the PL spectrum, which can be analyzed for the different isotopes and elements constituting a given PL centre. The availability of ^{28}Si proves that an early suggestion regarding the detrimental influence of the isotope distribution of the Si host material on the PL linewidths of IBE centres is correct [38].

1.1.4 The Experiment

Isotopically enriched ^{28}Si was used to examine isotopic fingerprints of many previously studied IBE centres and to describe a number of closely related, newly discovered centres. The reduced spectroscopic linewidth of ^{28}Si enabled this novel research method, which can deliver valuable new information concerning the actual composition of deep impurity centres. The comprehensive data presented here includes isotopic fingerprints of all studied centres, and the no-phonon (NP) PL energies and their isotope shifts. No-phonon transitions do not involve any vibrational modes. In addition, many of the observed NP PL lines show local and pseudo local vibrational mode (LVM/pLVM) replicas, which appear shifted in the optical spectrum by the energy amount of the involved vibrational mode. The energies and isotope shifts of these replicas are presented as well. These results provide valuable input for the much-needed theoretical studies and modelling of the impurity centres that will explain the formation and properties of these ubiquitous complexes. Initial modelling attempts have been hampered by the incorrect assignment of constituents based on earlier experimental results. While these early results were based on elaborate, well planned experiments, they suffered from low spectroscopic resolution, uncontrolled contaminants and, of course, a lack of pure, isotopically enriched material, especially ^{28}Si. On the basis of preliminarily published results of this research project, some modelling efforts for the prototypical 1014 meV Cu_4 centre have already begun [39–44].

Many of the IBE centres studied here have been extensively discussed in the literature, therefore, in Chap. 2 these earlier publications on experimental, as well as theoretical, results are reviewed. The experimental methods employed in this thesis, including a discussion about the sample preparation methods and Fourier transform spectroscopy, are introduced in Chap. 3. Finally, the most recent experimental results are presented in Chap. 4 and discussed in Sect. 5.1, while Sect. 5.2 provides the conclusion.

The following sections provide the necessary background information to understand and analyze the experimental results presented as part of this thesis. The concept of isoelectronic bound excitons (IBE) and their formation and characteristic properties are explained, followed by a look at impurity and host isotope effects and the local vibrational mode (LVM) replicas of IBE. Finally, the effects of transition metals (TM) in Si are examined, with special attention given to Cu.

1.2 Bound Excitons

The photoluminescence (PL) centres studied here are characterized as IBE. To better understand IBE, it is important to first look at bound excitons (BE). A BE is an electron-hole pair that is localized at an impurity or defect, as illustrated for a donor BE in Fig. 1.2. The BE can also be understood as a localized electronic excitation of the binding centre. Because the excitation is localized, all BE centres share one very important characteristic: their transitions are exceptionally sharp because the excitation has no net translational kinetic energy. This characteristic is in strong contrast to the free exciton (FE) or other intrinsic processes, which possess kinetic energy according to the Boltzmann distribution, thus having a PL lineshape which is a Maxwell-Boltzmann distribution. This property was recognized by Haynes [45] in the first report of PL due to BE in a semiconductor, namely Si, and further investigated by Dean et al. [46].

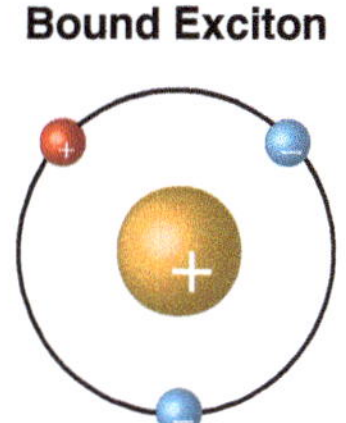
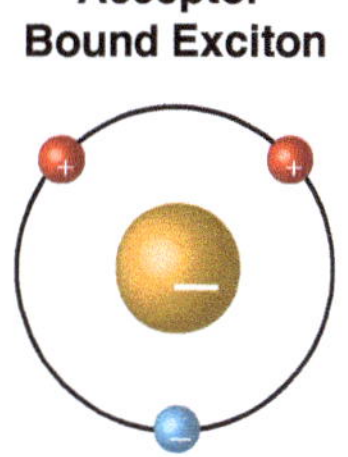

Fig. 1.2 Schematic diagram outlining the structure of donor and acceptor BE, respectively. The donor BE consists of a positive core with two bound electrons and one bound hole. The acceptor BE binds two holes and one electron to a negative core

These initially reported BE were not IBE, but rather BE associated with neutral shallow donor impurities such as As or P and shallow acceptor impurities such as B in Si, and they possess many properties that are quite different than those typical of IBE. The shallow donor and acceptor BE usually have a very small localization energy, which is the energy required to dissociate the BE into a neutral impurity and a FE with zero kinetic energy. Their PL is therefore just below the actual band gap energy of the host, with the no-phonon (NP) transition located at the band gap energy minus the FE binding energy and the BE localization energy. Furthermore, this energy closely tracks the band gap energy as a function of for example temperature or hydrostatic pressure. Due to the indirect band gap of Si and the requirement for wavevector conservation, the NP transition is weak, allowed only by scattering from the central cell potential of the impurity ion, since the localization arises mostly from the Coulomb interaction and the exchange/correlation energy of the mobile electronic particles. Much of the PL (or absorption) intensity of these shallow BE occurs with the participation of wavevector conserving phonons (WCP). This leads to phonon replicas where the wavevector difference between the top of the valence band and the bottom of a conduction band is accommodated by the emission (the only process possible at low temperature) of an appropriate phonon. In Si, the dominant WCP replica is the transverse optical (TO), but the longitudinal optical (LO) and transverse acoustic (TA) WCP also result in relatively strong replicas. Another consequence of the indirect band gap for the shallow donor and acceptor BE is the very low PL quantum efficiency that results from the long radiative lifetime, thus leading to the dominance of nonradiative Auger decay [47, 48]. Further details of shallow donor and acceptor BE in Si are found in several reviews [49, 50].

1.3 Isoelectronic Bound Excitons

An IBE is an electron-hole pair bound to a neutral, isoelectronic impurity in a semiconductor. The term originated from cases where the binding centre is a single isoelectronic substitution from the same column of the periodic table as the atom being replaced, therefore having the same number of valence electrons. Unlike donor or acceptor impurities, which come from different columns of the periodic table than the host atom they are replacing, these substitutions do not change the valence or bonding, therefore we would not expect "doping" effects from the release of weakly bound electrons or holes. Some of the first spectroscopic observations on such isoelectronic impurity centres were made by Dietz et al., Thomas et al., Trumbore et al. and Hopfield et al. [51–54]. These centres were characterized as isoelectronic traps where, for example, N substitutes for P in GaP (GaP:N_P) [52], or GaP:Bi_P [53], or ZnTe:O_{Te} [54]. Hopfield et al. [54] were also the first to distinguish between isoelectronic donors (GaP:Bi) and acceptors (ZnTe:O) due to the similarity of the excited state optical spectrum to shallow donors and acceptors. They are also referred to as donor- and acceptor-like IBE, or as isoelectronic pseudodonors and isoelectronic pseudoacceptors.

For Si, no such simple substitutional IBE exist. Substitutions with the group IV elements C, Ge, Sn, and Pb do not result in centres capable of localizing a carrier or exciton. This experimental fact was explained by Baldereschi et al. [55], who showed that the localization energies of such simple single substitutional IBE could be reasonably well explained by proper consideration of screening and lattice relaxation. They also predicted that unlike shallow donor and acceptor BE, the hydrostatic pressure shifts of IBE could be quite different from those of the band gap—a typical characteristic of deep centres. Eventually, it was discovered that the formation of IBE centres having complex, multi-atom cores in Si is possible and that they show the same characteristics of the aforementioned single-substitutional IBE. As a consequence of this added complexity, the structure of IBE centres in Si is typically more complicated and their symmetry is lower than cubic. Many of the early studies on IBE in Si give special attention to the determination of the binding centre's symmetry. The properties and details of IBE, with a particular focus on Si, are discussed in a number of reviews [5, 38, 56–60].

The first PL centre identified as an IBE in Si was described by Weber et al. [61]. The so-called A, B, C centre was first suggested to contain C [61], which was later revised [62] and evidence of N [63] and later Al-N [64] being part of this centre was presented. The A, B, C centre was shown to have trigonal, or C_{3v} symmetry, and to be an acceptor-like IBE [11, 65]. Almost at the same time, another PL system, the long-lived P, Q, R PL lines were reported in In-doped Si and attributed to In related IBE by Mitchard et al. [66]. Shortly thereafter, Weber et al. [67] and Thewalt et al. [68] independently discovered that the P, Q, R system could be greatly enhanced by rapidly quenching In-doped Si from ∼800 °C to room temperature. Weber et al. [67] suggested that the centre was an In-Fe pair. A similar system, with the same enhancement due to rapid thermal quenching, was then discovered in Tl-doped Si [69–72]. While Schlesinger et al. [70] first supported the incorporation of Fe in the In- and Tl-related IBE centres, this proposal was later cast into doubt when the same group failed to observe any isotope shift in the PL transitions when diffusing in different Fe isotopes [73]. To this day, the detailed constituents of the In- and Tl-related IBE centres remain unknown.

The discovery of the role of thermal quenching in the creation of IBE centres in Si [67, 68] has since played an important role in the study of these centres. This treatment facilitates the formation of impurity complexes that incorporate TM, as their high solubility and diffusivity at high temperatures would otherwise drive these impurities to the sample surface during a slow cool-down process. Rapid quenching "freezes in" a large supersaturation of interstitial TM into the Si bulk. Thermal quenching is necessary to obtain strong PL for most IBE centres studied here.

Soon after the discovery that the PL of the In-related IBE could be greatly enhanced by thermal quenching, an intense IBE system with NP lines at $\sim$1014 meV was reported by Weber et al. [9] and Watkins et al. [74], and ascribed by the former to recombination at a Cu-pair centre. The 1014 meV "Cu-pair" centre is prototypical of the TM-containing centres whose constituents have been re-evaluated by the isotopic fingerprint technique, which are the focus of this thesis. The many previous studies of TM-containing PL centres are discussed in greater detail in Chap. 2.

One of the few IBE centres in Si whose identity was firmly established before the advent of isotopic fingerprints was the Be-pair centre, first reported by Henry et al. [75]. This centre, in which two Be atoms replace a single Si atom, has been studied extensively. Most of the techniques that have been applied to IBE centres, including uniaxial stress and Zeeman spectroscopy, PL lifetime versus temperature studies, and spectroscopy of the IBE excited states have been used, establishing an acceptor-like character [75–86].

1.3.1 IBE Binding Mechanism

In Si, no simple isoelectronic group IV substitutions form IBE centres [55], but complexes of other elements can form an isoelectronic binding centre. In the ground state these centres introduce no additional charge to the system, but they may bind a charge carrier in their local central-cell potential, which has a short range character, leading to a high localization of the charge carrier. The central-cell potential is caused by local effects, such as differences in the local electronic bonds between the impurity defect centre and its lattice neighbours, compared to the character of the bonds in a perfect, undisturbed crystal. Hopfield et al. [54] first pointed out the possibility of creating two different centres—if the energy of the s orbital is lowered, the centre can attract an electron from the conduction band; if the energy is increased, a hole from the valence band can be bound. The centre then possesses a Coulomb field and can capture another particle of opposite charge, making it neutral again [58, 60]. A centre that first captures an electron and then a hole is called an acceptor-like IBE. If the centre captures a hole first, subsequently attracting an electron, then it is called a donor-like IBE [54]. The highly localized charge of the primary particle leads to a lattice relaxation, which, in turn, reduces the energy of the particle [5]. The electron-hole pair can then be described as an exciton localized at the isoelectronic impurity centre, hence the name isoelectronic bound exciton (IBE) centre. Figure 1.3 illustrates the differences between regular donors and donor/acceptor-like IBE.

In the case of a donor-like IBE centre, the core tightly binds a hole with a binding energy of a few hundred meV, and the resulting positively charged centre then binds an electron in a shallow effective mass orbit with a binding energy of much less than 0.1 eV [10], usually in the range of 30–40 meV [9].

The bound electron of a regular neutral donor experiences a long-range Coulomb potential already in the ground state due to the charged core. Defining the ground state of the IBE centre as having no electrons or holes bound to the neutral core, there is no Coulomb potential in the ground state of the IBE. The Coulomb potential only exists for the bound exciton state due to the charge that is tightly bound to the core [87]. Labelling the levels according to T_d symmetry, for a donor-like IBE only transitions from $1s(A_1)$ to the ground state of the IBE centre are observed in PL spectroscopy. $1s(A_1)$ forms the ground state of the BE. For a regular donor the ground state is the $1s(A_1)$ level, and observable transitions are between higher excited states and this ground state.

The excited states of the donor-like IBE are then effective mass approximation (EMA) states in the Coulomb field of the tightly bound hole, with the difference that, in this system, the electron spin can couple to the hole spin, whereas for a neutral donor the interaction with the positive core is taken to be only Coulombic. This spin coupling is the source of the splitting of the IBE ground state into a multiplet, as described in detail in Sect. 1.3.3. An alternative view of a donor-like IBE would be to think of it as a very deep neutral donor, effectively designating its ground state as the $1s(A_1)$ level.

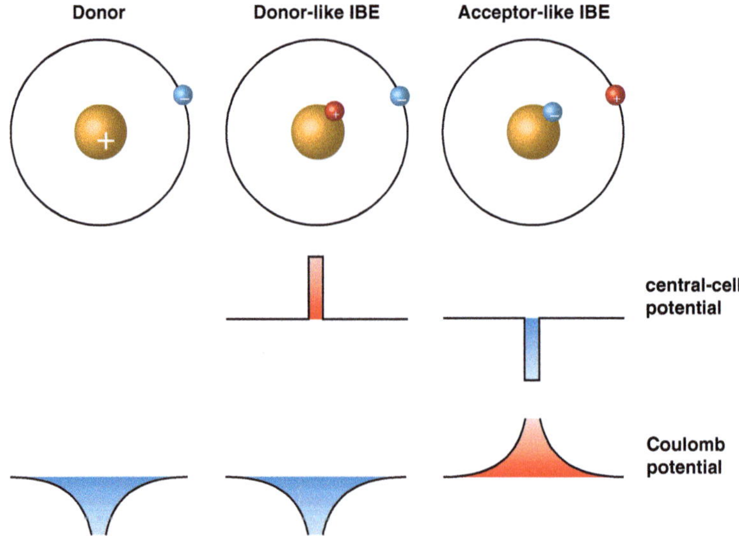

Fig. 1.3 Schematic diagram outlining the differences between a regular donor and a donor-like IBE. Also shown is a diagram for an acceptor-like IBE. The central-cell potential tightly binding the primary particle in isoelectronic centres is indicated, as well as the resulting Coulomb potential, which, in turn, attracts the secondary particle. The Coulomb potential due to the core charge is also shown for the regular donor [60]

To experimentally determine whether an IBE has a donor or acceptor character, the excited states of the Coulombically bound particle have to be studied. This can be done either with absorption or photoluminescence excitation (PLE) experiments. Photoluminescence experiments at elevated temperatures are difficult because the PL of IBE centres usually vanishes. The excited states that can be observed in these experiments are very similar in character to effective mass approximation (EMA) states of shallow impurities. The central-cell potential only causes minor changes to the extended EMA p-states of the impurity centre, but those s states that have a nonvanishing amplitude of their wavefunctions at the defect site are significantly altered by valley-orbit-interaction [88, 89]. This behaviour has been shown for the excited Coulomb states of other deep donors; for example, for S and Se related centres [22, 23, 90], and for deep acceptors such as Au and Pt [91–93].

1.3.2 Characteristic Properties of Isoelectronic Bound Excitons

For many regular shallow donor or acceptor defects, the BE recombination is usually weak or even nonradiative due to strong nonradiative Auger effects [47, 48]. The Auger effect causes the resulting BE recombination energy to be used to ionize the remaining donor electron (acceptor hole) into the conduction (valence) band, leaving behind an ionized impurity centre. The resulting low optical quantum efficiency makes PL studies difficult, but absorption spectroscopy experiments on these centres can still be useful.

In contrast, IBE centres do not suffer from Auger recombination because they have only a single electron and a single hole. The IBE lifetime is determined by the radiative recombination probability of the exciton [5, 59] because IBE have a near-unity quantum efficiency [5, 56, 60]. A further effect is the relaxation of the requirement for wavevector conservation, which results from the strong localization of one of the trapped carriers. Therefore the PL of IBE is characterized by strong NP lines and very weak or absent WCP replicas. In addition, IBE have a narrow electronic transition energy, reflected in a sharp line in optical spectra, which is an important property for the experimental results presented here.

In the ground state, when no exciton is bound, IBE centres introduce no net charge to the system, therefore they are also spinless [59]. The exciton binding energy or localization energy E_{BE} can be determined through the band gap energy E_{G}, the energy of the free exciton E_{FE} and the measured NP photon energy $h\nu$

(Ref. [59], values for ^{nat}Si):

$$E_{\text{BE}} = E_{\text{G}} - E_{\text{FE}} - h\nu$$
$$= 1169.9\,\text{meV} - 14.7\,\text{meV} - h\nu$$
$$= 1155.2\,\text{meV} - h\nu$$

1.3.3 Electronic Levels of Isoelectronic Bound Exciton Centres

The electronic level scheme of the ground states of donor-like IBE centres is a result of the electron-hole coupling of the spin of the hydrogenic electron and the tightly bound hole [59]. Through their Coulomb fields the $S = 1/2$ electrons and the $S = 3/2$ holes are j-j-coupled, splitting the energy level into a 3-fold degenerate $J = 1$ and a 5-fold degenerate $J = 2$ level with the separation being the j-j exchange energy [94]. In addition to this effect, lattice distortions introduced by the presence of the defect centre partially lift the degeneracies of electron and hole states [94]. This internal strain splits the $m_{\text{h}} = \pm 3/2$ hole states from the $m_{\text{h}} = \pm 1/2$ hole states. In the presence of both j-j interaction and internal strain splitting and in absence of any external strain or magnetic fields, the exciton ground state manifold can split into a total of up to five levels $|J, M\rangle$, where M is the z-component of J:

$$|1, 0\rangle, \ |1, \pm 1\rangle \tag{1.1}$$
$$|2, 0\rangle, \ |2, \pm 1\rangle, \ |2, \pm 2 \tag{1.2}$$

These ground state splittings of the exciton can range from sub-meV to several meV in Si. Transitions from IBE states having $J = 2$ are forbidden in the dipole approximation. Luminescence from such forbidden transitions is usually extremely weak, but still visible due to higher order effects. This can result in dramatic changes in the PL spectrum and PL lifetime even between 4.2 K and lower temperatures, due to thermalization between the ground state components. This is particularly true if the lowest energy ground state component is forbidden, in which case the NP PL can essentially disappear in the limit of low temperature, leaving only the phonon and local mode replicas. An example of this is the 1066 meV Cu_4Au centre, which is detailed in Sect. 4.2.3 [95].

The actual symmetry of a given defect centre can often be determined in Zeeman or uniaxial stress experiments, but the observed symmetry in PL can be lower than the actual symmetry of the complex due to lattice distortions caused by the IBE [5, 96].

1.4 Isotope Effects

The IBE centres studied here using PL display isotope effects due to two different sources. One source is the isotopic constitution of the Si host crystal, which has a number of different effects. Hence the isotopic enrichment of the samples used for the experiments discussed here plays a major role in the results of this project. The second source is the varying isotopic mass of the constituents of the IBE centre, causing an isotope shift in the observable PL signal, which is ultimately responsible for the observed isotopic fingerprint fine structure.

1.4.1 Host Isotope Effects

Before many host isotope effects were discovered with the help of isotopically enriched ^{28}Si, many properties of semiconductors and their defects, such as their band gap or impurity transition energies, had been explained using the virtual crystal approximation (VCA) [24]. In the VCA, a real crystal that is composed of several different isotopes is approximated by a perfect crystal with a constant mass throughout the lattice sites. This mass is the average isotope mass of the actual atoms. As a consequence, in the VCA a crystal made of a single isotope would have the same properties as a crystal that is composed of several different isotopes with the same average isotopic mass. The VCA was thought to be an adequate approximation, as expected isotope effects would be small and below detection limits. Only later experiments with highly enriched ^{28}Si revealed that many effects that were originally thought to be intrinsic or fundamental limits of a given system, were indeed a result of the isotopic randomness of ^{nat}Si [24]. Optical studies of isotopically enriched ^{28}Si have given rise to a number of unforeseen results, due to the significantly reduced spectroscopic linewidths that result from the elimination of inhomogeneous isotope broadening [15, 16]. As mentioned in the introduction, these results range from shallow [16–21] and deep [22, 23] IR absorption spectroscopy over BE PL [15, 19, 24–26] and PLE [19, 27–29, 31] spectroscopy, to the deep IBE PL spectroscopy discussed here [32–37].

1.4.1.1 Inhomogeneous Isotope Broadening

One goal of optical defect spectroscopy has always been to obtain the narrowest possible optical linewidth, as this opens the possibility to study defects at a much higher resolution and to observe otherwise hidden fine structure. In early spectroscopic studies of impurity centres in Si, the widths of optical absorption or emission lines resulting from transitions between the electronic ground state and a bound excited state of the impurity centre (or vice versa) were dominated by inhomogeneous fields and inter-impurity interactions as a consequence of high impurity concentrations and imperfect crystals. Progress was made in controlling the crystalline perfection and the chemical purity of Si single crystals, reducing the linewidth of optical transitions. The quality of the Si samples came to a point where no significant improvement in linewidth could be observed. This fact suggested that a fundamental lifetime broadening limit had been found and that all inhomogeneous broadening effects were eliminated. Barrie and Nishikawa [97, 98] proposed that this lifetime broadening was a result of phonon-assisted transitions from the excited state to other near-lying states. This mechanism was in reasonable agreement with the observed linewidths in high quality Si samples until recently, and therefore was assumed to be the limiting factor for the achievable linewidth [99]. The possibility of inhomogeneous isotope broadening in Si was mentioned very early in a review by Davies [38], but high purity, highly enriched Si was not available until much later. The advent of isotopically enriched Si immediately showed that the isotopic composition and isotopic randomness of the host material had a strong influence on the observed linewidth of optical absorption and emission transitions. Karaiskaj et al. [15] reported the first high resolution PL study of isotopically pure ^{28}Si. Subsequent studies have made it clear that the dominant broadening mechanism for many impurity absorption lines in ^{nat}Si is not the lifetime broadening as had been thought. Rather the inhomogeneous broadening is a result of the isotopic randomness present in ^{nat}Si [15, 16, 19, 24, 100]. It has been shown that isotopically enriched Si is key to achieving narrower optical transition linewidths than what was possible before.

The inhomogeneous broadening of PL transitions has been shown to result from local fluctuations in the band gap energy [15]. The dependence of the band gap energy on the isotopic composition of the crystal has been shown before for C (diamond) [101], Ge [102] and Si [99]. It was explained to result primarily from the effect of the average isotopic mass on the electron-phonon interaction, with a smaller contribution from the change in lattice constant. The resulting effect of statistical fluctuations of the band gap energy in the effective volume of an exciton leads to inhomogeneous broadening of the PL line. The removal of this inhomogeneous broadening has been observed for the first time in a semiconductor for shallow impurities in ^{28}Si [15]. The observed reduction in linewidth was found in accordance with an effective excitonic volume of radius $\sim$3.5 nm.

After removing the inhomogeneous broadening by using isotopically enriched host material, ^{28}Si in this case, the observed optical linewidths are much narrower. If one were able to remove all inhomogeneous broadening, only homogeneous broadening due to the actual lifetime of the excited state would be left, indicated by a Lorentzian line shape. The lower limit to the observed linewidth is dictated by the uncertainty principle $\Delta E \, \tau = \hbar/2$. The longer the lifetime τ, the smaller the energy uncertainty ΔE and the full-width at half-maximum (FWHM). The lifetimes of IBE are very long (see for example Sect. 2.1), and the observed linewidths clearly show that the FWHM of IBE PL transitions are not lifetime limited. After removing the inhomogeneous isotope broadening, effects such as random electrical fields or strains continue to contribute to the observed linewidth.

1.4.1.2 Other Host Isotope Effects

By using samples of different isotopic Si compositions, the effect of the isotopic composition of the crystal on the indirect band gap of Si was shown [17–19]. The shift between ^{nat}Si and ^{28}Si was determined to be 114 µeV, with ^{28}Si having the smaller gap energy. Consequently, this effect can be observed in the NP transitions reported here as well, albeit to a smaller extent due to the deep character of the studied IBE centres. For example, in ^{28}Si the NP transition energy of Cu$_4$ is downshifted by about 62 µeV from ^{nat}Si.

Other host isotope effects include the removal of the "intrinsic" acceptor ground state splitting in ^{28}Si [24, 103]. The influence of the average atomic mass of the host on the host phonon frequencies and isotopic randomness can have more subtle effects on the phonon spectrum [17], especially the thermal conductivity [19]. However, since WCP replicas do not play a major role in the study of IBE transitions, there is little consequence on the spectroscopic results of these centres.

The increased sharpness of shallow BE transitions in ^{28}Si has also enabled the observation of the temperature dependence of the bandgap E_G of Si for temperatures $T < 4.8\,K$ [24, 104]. It was found that $E_G \propto T^4$ for $T \rightarrow 0\,K$, in agreement with theory [104].

1.4.2 Impurity Isotope Shift

Isotopic substitutions within a given impurity centre can lead to a shift in the NP lines of IBE in optical spectra, as well as of their LVM and pseudo LVM (pLVM) replicas. These shifts can be of variable size depending on the impurity mass, the difference in isotope masses, and the complex itself. Typically, these shifts are very small and they can only be resolved in high quality samples with the highest spectroscopic resolution. Before highly purified ^{28}Si was available for optical spectroscopy experiments, this shift could be observed in ^{nat}Si [9]; however, because the isotope shift is almost always smaller than the PL linewidth in ^{nat}Si, only a simple shift of the NP line could be observed. While this would confirm the presence of that element in the binding centre, no conclusion could be made about the number of atoms of that element in the binding centre. One notable exception to this limitation of ^{nat}Si is the lithium vacancy complex Li$_4$ V$_{Si}$, where isotopic fingerprints could be resolved in natural Si [38, 105–107]. This complex was thought to be a special case, since the very light Li atoms resulted in extremely large NP isotope shifts.

A quantitative explanation of the isotope shift of NP transitions due to electronic transitions at deep centres was given by Heine et al. [108]. NP transitions do not involve WCP, but both, the initial state and the final state include the zero-point energy of all the lattice vibrations $E_0 = \sum_i 1/2\,\hbar\omega_i$. A change in the state of the electronic system causes a change in the force constants between the atoms, and, therefore in the vibrational frequency ω_i. For example, a hole weakens the covalent bonds in a tetrahedral lattice and lowers ω_i. This is normally part of the total transition energy, and not observable separately. Similarly, an isotopic substitution of a defect centre also causes a finite change in zero point energy due to the mass change, however, because it is the same for initial and final states, it is also unobservable. But there is a cross-term between the mode softening caused by binding a carrier (electron-phonon interaction) and the isotopic substitution, which is measurable and constitutes the isotope shift that is observed in optical spectra of the NP lines [108, 109].

In a simple description of this isotope shift [108], we consider an oscillator with a force constant Λ and $M\omega^2 = \Lambda$. Suppose the presence of the carrier lowers Λ by a fraction of γ_c through the electron-phonon interaction. Because the electron wave function is spread over a number of atoms in the lattice, the force constant is softened with a probability P.

$$\omega^2 = \frac{\Lambda}{M}\left(1 - \gamma_c P\right) \tag{1.3}$$

Now we substitute the mass M with $M + \Delta M$ and expand to the lowest power of ΔM and $\Delta\omega$ [108] and obtain:

$$\omega^2 + 2\omega\Delta\omega = \Lambda\left(\frac{1}{M} - \frac{\Delta M}{M^2}\right)\left(1 - \gamma_c P\right) \tag{1.4}$$

$$= \frac{\Lambda}{M}\left[1 - \underbrace{\gamma_c P}_{\substack{\text{mode} \\ \text{softening}}} - \underbrace{\frac{\Delta M}{M}}_{\substack{\text{isotopic} \\ \text{substitution}}} + \underbrace{\gamma_c P \frac{\Delta M}{M}}_{\substack{\text{isotope shift} \\ \text{cross term}}}\right] \tag{1.5}$$

The resulting terms are the mode softening, the isotopic substitution and the isotope shift cross term. Hence, the observed change in frequency due to the isotope shift cross term is then the cross term itself [108]

$$\Delta\omega = \frac{1}{2\omega}\frac{\Lambda}{M}\frac{\Delta M}{M}\gamma_c P. \tag{1.6}$$

This makes the isotope shift S for all three modes of vibration

$$S = \frac{3}{2}\hbar\Delta\omega \tag{1.7}$$

$$= \frac{3}{4}\frac{\hbar}{\omega}\frac{\Lambda}{M}\frac{\Delta M}{M}\gamma_c P. \tag{1.8}$$

And for $\gamma_c P \ll 1$ we obtain

$$S \approx \frac{3}{4}\hbar\omega\frac{\Delta M}{M}\gamma_c P. \tag{1.9}$$

The isotope shift S depends inversely on the actual mass M so that isotopic substitutions of lighter atoms have larger isotope shifts.

P is the probability for a bound carrier to be in the vicinity of the impurity and its nearest neighbour bonds. For regular donors and acceptors, this probability is relatively small due to the EMA-like states of the carrier, but P is much larger for isoelectronic traps where the primary carrier is tightly bound in the central-cell potential. Naturally, in both cases the probability increases with increased binding energy. As a result, there is no sizeable shift for strictly hydrogenic centres [108], with a notable exception being B in Si, where the observed isotope shift is unusually large [16, 21]. Generally, however, the more the central-cell correction pulls the wave function into the core, the larger the observable shift. So for deep centres, such as those discussed here, the wave function is localized enough to result in a significant probability P [108] and, therefore, a significant isotope shift of the optical transitions can be observed.

1.5 Local Vibrational Mode Replicas

Due to the character of IBE, the need for wavevector conserving phonons during the electron-hole recombination is lifted, but usually not the entire PL intensity is contained in the NP transition. Most IBE centres have local vibrational mode replicas.

Lattice relaxation plays an important role in the IBE binding mechanism because it can bring with it strong vibronic coupling effects, as studied in detail for the *A, B, C* centre by Davies et al. [110]. Further details of vibronic effects can be found in the reviews by Davies et al. [38, 111]. In some cases, the vibronic sideband of an IBE transition can map out the entire phonon density of states for the host Si lattice. However for other centres, the vibronic coupling can be dominated by vibrational excitations of the binding centre itself. These excitations are often referred to as local vibrational mode (LVM) replicas, although, strictly speaking, a true local mode must not be resonant with the host lattice modes. In the case of Si this means that it must have an energy larger than $\sim$65 meV, resulting either from the vibration of a constituent lighter than Si or from an increased bond strength. Here the term LVM is reserved for such true local modes, an example of which is the 104.7 meV mode seen for the Be-pair centre by Henry et al. [78]. Other such LVMs are discussed in Chap. 4 in connection with TM-containing IBE that also contain Li, a constituent with a very small mass.

Most IBE centres display a very characteristic spectrum of low energy satellites [59], which are vibrational replicas of the NP line. These satellites are not wavevector conserving crystal modes or high energy LVM, but in-band resonant modes due to local vibrations of the impurity centre in resonance with the Si lattice modes. They have relatively well defined energies and we refer to them as pLVM. Most of the TM-containing IBE centres that are discussed here display very characteristic spectra of pLVM replicas, with energies in the range of 6 to 11 meV. It should be emphasized that these characteristic pLVM replicas of the TM-containing IBE previously observed in ^{nat}Si samples were also observed in the same ^{28}Si samples from which the isotopic fingerprint results were obtained, thus confirming that the same centres were being studied.

It is generally not known which atomic displacements are involved in these pLVM modes, which makes the calculation of mode energies from first principles difficult [112]. For both kinds of vibrational mode replicas, the mode energy depends on the constituents of the luminescent impurity centre, hence the energy changes with the constituents' isotopic masses [108].

The energies of these LVM and pLVM replicas can provide an important test for models of the binding centres. As discussed in Chap. 2, Estreicher and coworkers have recently calculated the stability and vibrational properties of many Cu-containing complexes to provide detailed models for the structure of these complexes [112–116]. An early example of such modelling is the *ab-initio* local-density calculation of the stability and vibrational properties of various Be-containing defects by Tarnow et al. [85], who predicted a [111] oriented substitutional-interstitial pair as the lowest energy configuration of a Be-pair in Si and further predicted the LVM energy for this configuration. The observation of an LVM replica for the Be-pair IBE centre at almost exactly this predicted energy by Henry et al. [78] provided strong confirmation of the substitutional-interstitial configuration of the defect.

1.5.1 Isotope Effects on Vibrational Mode Replicas

Isotopic substitutions of the binding centre constituents can have a relatively large effect on the LVM and pLVM replicas associated with these IBE. In a simple approximation, the energy of a vibrational mode is given by the proportionality

$$E_p \propto \frac{1}{\sqrt{M}}, \qquad (1.10)$$

where M is the mass of the vibrating body. Thus it is expected that the vibrational mode energy E_p changes with the isotopic composition of the IBE centre if a given mode primarily involves the motion of one constituent atomic species. These shifts can be measured with respect to the NP line. However, the heavier the vibrating mass, the smaller the observable difference in energy for different isotopes. Also, vibrational mode replicas have an inherent spectroscopic linewidth. Thus the possibility of observing these relatively large shifts (as compared to NP shifts) is usually limited by signal-to-noise considerations and other limiting factors, like overlapping nearby modes.

This effect has indeed been used in the past to prove the presence of specific elements in an IBE binding centre. The pLVM range in energy between 6 and 11 meV for the IBE centres studied here, and they can usually be attributed to heavy impurity atoms. In literature, the observed isotope shifts are often quoted as ratios

$$r_E = \frac{E_p(I_l)}{E_p(I_h)},\tag{1.11}$$

where E_p is the vibrational mode energy, and I_l and I_h symbolize the lighter and heavier isotopes of a given impurity element, respectively. Due to the inverse proportionality of E_p and M, these ratios can be compared to the square root of the ratio of the heavier isotope's mass number over the lighter one

$$r_m = \sqrt{\frac{m(I_h)}{m(I_l)}}.\tag{1.12}$$

Table 1.1 lists r_m for some of the elements used in this study.

1.6 Transition Metals in Si

Transition metal impurities in Si exist in different charge states—they can occupy substitutional and interstitial lattice sites, and they can form complexes with themselves and with other impurities [11]. In this section, some of the TM used here are introduced before some general effects of TM on Si are discussed. Finally, the solubility and diffusivity of Cu in Si are reviewed in greater detail, because Cu is central to the formation of almost all IBE centres discussed in this work.

Cu The TM Cu has three characteristic properties: It always diffuses interstitially in a positive charge state, only has a weak interaction with the Si lattice, and, upon the formation of silicides, a large lattice expansion is observed. Together, these properties are mainly responsible for the behavior of Cu in Si [6]. The optical centres discussed here form at relatively low Cu concentrations, and precipitation is generally avoided in this research. Cu is fairly ubiquitous and is a very fast diffuser in Si, therefore it is problematic in Si processing steps that involve elevated temperatures. Substitutional Cu forms a deep impurity centre in Si, with energy levels near the middle of the band gap. Hence the

Table 1.1 In a simple approximation, the ratio $r_E = E_p(I_l)/E_p(I_h)$ of the energies of LVM of light (I_l) and heavy (I_h) impurity isotopes is expected to behave inversely to the ratio $r_m = \sqrt{m(I_h)/m(I_l)}$

I_h	I_l	r_m
^{65}Cu	^{63}Cu	1.016
^{109}Ag	^{107}Ag	1.009
^{197}Au	^{195}Au	1.005

The ratios r_m for the isotopes used in these studies are given here conductivity and free carrier lifetime of the Si semiconductor is modified [9]. Many deep impurities form complexes with other impurities and it was thought that Cu would do the same. After the initial discoveries of the role of Cu in the formation of defects in Si, Cu has started to play an important role for gaining understanding of the many aspects of fundamental defect physics [117].

Pt As an unintentional contaminant, Pt usually poses a minor problem, as it is not very prevalent in the environment [4]. However, with its high carrier capture cross section, Pt is commonly used in device engineering to decrease the minority carrier lifetime to increase the switching speed of silicon devices [4]. At high temperatures, Pt is a quick interstitial diffuser in Si, but Pt can also occupy substitutional sites at a much slower rate via a kick-out mechanism, leaving behind Si self interstitials [4]. The solubility of interstitial Pt in Si is low, and after a suitable diffusion process at 800–900 °C the concentration of substitutional Pt surpasses that of interstitial Pt. As a consequence, low temperature annealing does not alter the Pt content of a Si sample significantly [4].

Au In Si, Au is a strong recombination centre and concentrations of much less than $10^{12}\,\mathrm{cm}^{-3}$ impact device performance [4]. The diffusion behavior of Au is very similar to that of Pt, but the required diffusion temperature is lower (500 °C). After cooling the sample, Au remains on substitutional lattice sites, which it occupies through the kick-out process [4].

Ag As a fast diffuser, Ag is assumed to diffuse interstitially. Because of the similarity of Au and Ag donor and acceptor levels after high temperature diffusion, it is assumed that Ag forms substitutional defects [4]. As with Pt and Au, defect concentrations are found to be much higher for Si samples that contain dislocations.

1.6.1 Effect of Transition Metals on Si

The effects of TM on Si were an early research topic and Hopkins et al. [2] and Rohatgi et al. [3] recognized the deleterious effect of Ti, Fe, and Cu on the efficiency of Si solar cells. A further review of TM in Si was published by Weber [118]. The effects of Cu in Si have been reviewed by Istratov et al. [6], which included discussions on the diffusivity and defect reactions of Cu in Si, the impact on Si devices, and the development of Cu diffusion barriers. The characteristics of Cu diffusion barriers are especially important when Cu is used in large scale Si integrated circuits to form interconnections, as is standard technology today [119]. Istratov et al. [120] also reviewed the effects of iron contamination in Si. Cu is detrimental for Si devices even at trace contaminations below $1 \times 10^{12}\,\mathrm{cm}^{-3}$ [12]. Hall et al. [121] described the diffusion process of Cu in Si, which was later extended by Istratov et al. [122]. See Sect. 1.6.3 for a more detailed discussion of the Cu diffusion process. A comprehensive review of the influence of many metal impurities on Si and Si devices was written by Graff [4].

In general, Si device characteristics are altered by TM in several different ways. The deep donor and acceptor energy levels introduced by TM impurities affect the doping level of Si samples. A change in doping level can occur at a TM content of only a few percent of the original doping concentration of the material. Today this unintentional doping is well controlled and poses a problem only in high resistivity (low doping) Si devices. These contaminants diffuse to the surface during the low temperature processes that typically follow the high temperature diffusion treatments during device fabrication [4].

A more important factor affecting the Si device performance is the minority-carrier capture cross-section of TM traps, as a high capture cross-section drastically reduces the minority-carrier lifetime in the sample [4]. The capture cross-section for different impurities varies widely over orders of magnitude. Thus, only a small amount of an impurity with a high cross-section can alter the lifetime significantly. Moreover, the capture cross-sections for electrons or holes can be different by orders of magnitude as well. Even today, there is a lack of comprehensive and reliable experimental data for the carrier capture cross-section of many TM. Often, only the capture cross-section of the majority carriers has been measured, but these results vary by orders of magnitude [4].

A further important factor influencing Si devices is the generation leakage current [4]. This leakage current depends on the energy level, and the concentration of the trap and the carrier lifetimes. The carrier lifetimes are a function of the capture cross-section of the trap, as discussed before. Hence, different TM can have the same effect on carrier lifetimes, but the resulting leakage current can be vastly different. Both of these effects can be unwanted, but lifetime adjustments can also be necessary. For example, gold and platinum are often used to decrease the carrier lifetime of high voltage thyristors to improve the switching time, at the expense of an increased leakage current [123].

Another unwanted effect can be the precipitation of TM in Si at high concentrations. It was found in the 1960s that Cu in p-n junctions increases leakage current. At first, the mechanism of this degradation was not understood. The phenomenon was later explained by the formation of Cu precipitates, but the threshold Cu concentration is still unclear [6]. Precipitates within the space charge region of a p-n junction lead to a weak reverse current-voltage characteristic. The precipitate reduces the breakdown voltage and increases the leakage current across the junction. This effect can even be used to detect precipitates in a Si sample [4]. Precipitates also decrease the minority carrier lifetime, as they form a very efficient recombination channel [6].

It was found that the breakdown rate of oxide capacitors in metal oxide semiconductor (MOS) structures increases in the presence of Cu, starting with surface concentrations as low as $10^{10}\,\mathrm{cm}^{-2}$, depending on the oxide thickness. However, there is great disagreement among the different data presented in the literature [6], which may be attributed to the presence of suitable gettering sites for Cu within the Si material. Intentionally introduced gettering sites can improve the device yield. A suggested degradation mechanism is the formation of Cu-silicides at the Si–SiO$_2$ interface [6].

To fabricate integrated circuits that use Cu instead of Al interconnections, special diffusion barriers had to be developed to prevent the indiffusion of Cu from the interconnects into the bulk Si. Suitable barrier materials have sufficiently high conductivity, a very low Cu diffusion coefficient and good long term stability [6]. The use of Cu as a conductor is desired because of its high conductivity when compared to other, slowly diffusing metals like Al.

During many semiconductor processing steps, a small Cu contamination of the Si wafer is not a serious problem, as the material would undergo many heat treatments with a subsequent slow cooling process. This slow cooling allows the interstitial Cu to out-diffuse [4]. Only during rapid quenching procedures is the interstitially diffusing Cu_i locked into the crystal. To produce the IBE centres studied here, a rapid quench was often necessary to trap Cu_i and produce the impurity complexes in a sufficiently high concentration [68].

1.6.2 Solubility of Cu in Si

Hall et al. [121] had measured the solubility of Cu in Si in an early experiment with a radioisotope method using ^{64}Cu. However, Graff et al. [124] found the amount of Cu that is electrically active was much smaller than what had initially been determined. The concentration was found to peak at a diffusion temperature of only 800 °C. At these high temperatures, as much as 10^{18} cm^{-3} Cu can be dissolved in Si, yet the solubility at lower temperatures is vanishingly low and can go down to as little as $\sim$1 Cu atom per cm^3 in a room temperature equilibrium condition [6]. Furthermore, the amount of electrically active Cu also depends on the sample preparation method and the rate of cooling after the diffusion process. It was found that a fast quenching process is necessary to lock in and trap interstitial Cu_i in the Si crystal, which would otherwise diffuse out to the surface during a slow cooling process. This behavior was also observed for other IBE centres like Si:In [67, 68], where a fast quenching process increases the PL intensity of the centre by creating a supersaturation of impurities. During fast quenching, the "frozen in" Cu_i often forms complexes with other defect centres like Cu_s or other immobile impurities. Due to the small diffusion barrier and depending on the binding energy of the complex, these complexes can dissociate relatively easily and do not permanently trap Cu_i [6, 117]. At the highest contamination levels, Cu precipitation is a dominating process [117]. During slow cooling, most Cu diffuses to the surface because the small diffusion barrier of Cu_i allows for fast diffusion, but it can also decorate already present defects. A difference between the amount of Cu measured with the radioisotope tracer method [121] and the electrically active amount measured for example, by deep level transient spectroscopy (DLTS) [124] usually indicates the presence of Cu precipitates.

1.6.3 Diffusivity of Cu in Si

In many semiconducting crystals, among them Si, Cu is a fast diffuser [121]. In all cases, Cu diffuses interstitially as a singly positively charged Cu_i ion. In contrast to this, substitutional Cu_s is comparatively immobile because a Si vacancy is needed for its formation.

Reiss et al. [125] showed that the effective diffusivity can be lower than expected due to donor-acceptor pairing in the crystal. In this case, the Cu_i^+ can pair with the shallow acceptors B_s^- during the diffusion process. This trapping of Cu ions by B restricts the mobility of a large portion of the available Cu, which can then no longer contribute to the diffusion process. Hence, the measured coefficient is only the effective diffusion coefficient, which depends on the B concentration. A number of publications describe the Cu diffusion coefficient in Si (for example [126]) and their results were confirmed by Hall et al. [121], who assumed negligible Cu-B pairing in a highly doped p^+-Si sample ($\langle B \rangle = 5 \times 10^{20}$ cm^{-3}). A number of later publications attempted to correct Hall et al.'s coefficient for donor-acceptor pairing [127–130], but the work was done on the assumption of a strictly electronic binding of the donor-acceptor pair. It was only later shown that the dissociation energy of donor-acceptor pairs depends on the acceptor and, thus, has a covalent character [131]. As a result, these models had to be reconsidered [122].

In a recent review, Istratov et al. [6, 122] determined the intrinsic diffusion coefficient of Cu in Si for the first time to be:

$$D_{\text{int}} = 3.0 \times 10^{-4} \times e^{-0.18\,\text{eV}/k_B T}\ \text{cm}^2\,\text{s}^{-1} \tag{1.13}$$

Hall et al.'s [121] observed activation energy of the diffusion process was generally accepted as correct at the time, however it was much larger than the theoretical prediction of Woon et al. [132]. Istratov et al. measured an activation energy of 0.18 eV by using a low doping level p-type sample. The measured activation energy was close to the value predicted by Woon et al. These results were achieved in a sample with relatively low boron doping levels of 1.5×10^{14} cm^{-3} and at elevated temperatures, thus allowing the determination of the intrinsic diffusion coefficient directly from experimental data rather than compensating for doping levels later on. This method avoids the problematic donor-acceptor pairing of Cu_i^+ and B_s^- during the diffusion. Istratov et al. [122] arrived at the final expression to model the effective Cu diffusion coefficient at moderate doping levels in p-type Si material ($N_B \leq 10^{17}$ cm^{-3}):

$$D_{\text{eff}} = \frac{3 \times 10^{-4} \times e^{-2090/T}}{1 + 2.584 \times 10^{-20} \times e^{4990/T} \times (N_B/T)} \tag{1.14}$$

For heavily doped Si, a system of equations has to be solved [122]. The previously determined coefficient [121] remains valid as long as it is appropriately corrected for Cu-acceptor pairing. On the other hand, no experimental data is available for n-type

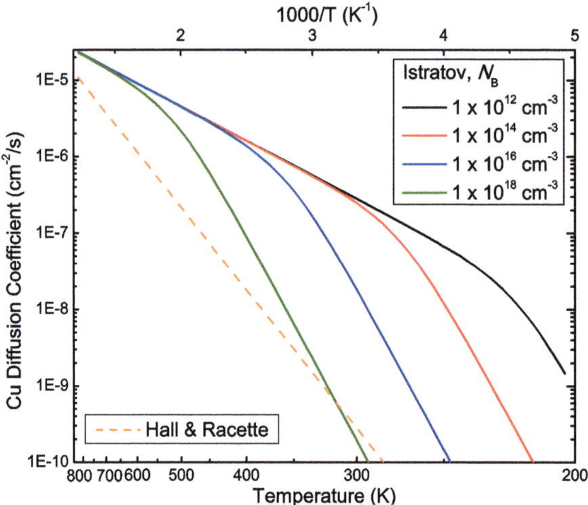

Fig. 1.4 Effective Cu diffusion coefficient according to Eq. (1.14) [6] for different acceptor concentrations N_B. At higher temperatures, the acceptor pairing becomes negligible. For comparison, the older Cu diffusion coefficient determined by Hall et al. [121] is also plotted

material, but it is thought that the intrinsic coefficient applies for moderately doped n-type material as well, because interstitial Cu_i^+ has never been shown to pair with shallow donors such as P [6]. D_{int} at room temperature is found to be three orders of magnitude larger than determined by Hall et al. [121], which means that Cu can diffuse through significant distances even at room temperature. The low activation energy of 0.18 eV is due mainly to the small ionic radius of Cu_i^+ in Si. In the absence of trapping centres, the Cu diffusion coefficient is large enough to enable not only the out-diffusion of Cu from a Si sample, but also the phenomenon of Cu contamination during chemomechanical polishing [133], which is, in principle, just a reversed out-diffusion. Nevertheless, at room temperature, and certainly at lower temperatures, the diffusion of Cu_i is often slowed down significantly by the formation of complexes with Cu_s or other impurities, as our long storage times of certain Cu diffused samples have shown.

Figure 1.4 shows the effective Cu diffusion coefficient of Eq. (1.14) plotted for three different boron acceptor concentrations N_B. Our ^{28}Si samples are p-type with N_B close to $10^{14}\,cm^{-3}$. At 800 K, the diffusion coefficient is less than $10^{-5}\,cm^2/s$, which is about 15 orders of magnitude larger than the coefficient of Al in Si [6], which diffuses substitutionally. Al is used traditionally to form interconnects in integrated circuits. This comparison shows that contamination of Si with Al is a negligible problem when compared to Cu contamination.

References

1. R. Landolt-Börnstein, *Numerical Data and Functional Relationships in Science and Technology, New Series, group III*, vol. 22. Semiconductors (Springer, Berlin, 1989)
2. R.H. Hopkins, R.G. Seidensticker, J.R. Davis, P. Rai-Choudhury, P.D. Blais, J.R. McCormick, Crystal growth considerations in the use of solar grade silicon. J. Cryst. Growth **42**, 493–498 (1977)
3. A. Rohatgi, J.R. Davis, R.H. Hopkins, P. Rai-Choudhury, P.G. McMullin, J.R. McCormick, Effect of titanium, copper and iron on silicon solar cells. Solid-State Electron. **23**, 415–422 (1980)
4. K. Graff, *Metal Impurities in Silicon-Device Fabrication* (Springer, Berlin, 2000)
5. G. Davies, Configurational instabilities at isoelectronic centres in silicon. Phys. Scr. **T54**, 7 (1994)
6. A.A. Istratov, E.R. Weber, Physics of copper in silicon. J. Electrochem. Soc. **149**, G21–G30 (2002)
7. T.G. Brown, D.G. Hall, Optical emission at 1.32 μm from sulfur-doped crystalline silicon. Appl. Phys. Lett. **49**, 245–247 (1986)
8. N. Minaev, A. Mudryi, V. Tkachev, Radiative recombination at thermal defects in silicon. Sov. Phys. Semicond. **13**, 233 (1979)
9. J. Weber, H. Bauch, R. Sauer, Optical properties of copper in silicon: Excitons bound to isoelectronic copper pairs. Phys. Rev. B **25**, 7688–7699 (1982)
10. M. Singh, W. Chen, N. Son, B. Monemar, E. Janzén, Shallow excited states of deep luminescent centers in silicon. Solid State Commun. **93**, 415–418 (1995)
11. J. Weber, P. Wagner, Photoluminescence from deep states associated with Iron in silicon. J. Phys. Soc. Jpn. 49, Suppl. A, 263–266 (1980)
12. M. Nakamura, S. Ishiwari, A. Tanaka, Number of Cu atom(s) in the 1.014 eV photoluminescence copper center and the center's model in silicon crystal. Appl. Phys. Lett. **73**, 2325–2327 (1998)
13. M. Henry, S. Daly, C. Frehill, E. McGlynn, C. McDonagh, in *A Photoluminescence Study of Gold- and Platinum-Related Defects in Silicon Using Radioactive Transformations*, ed. by M. Scheffler, R. Zimmermann. Physics of Semiconductors: 23rd International Conference on the Physics of Semiconductors-ICPS, pp. 2713–2716 (1996)
14. H.B. Erzgräber, K. Schmalz, Correlation between the Cu-related luminescent center and a deep level in silicon. J. Appl. Phys. **78**, 4066–4068 (1995)
15. D. Karaiskaj, M. Thewalt, T. Ruf, M. Cardona, H.-J. Pohl, G. Deviatych, P. Sennikov, H. Riemann, Photoluminescence of isotopically purified silicon: how sharp are bound exciton transitions? Phys. Rev. Lett. **86**, 6010–6013 (2001)
16. D. Karaiskaj, J. Stotz, T. Meyer, M. Thewalt, M. Cardona, Impurity absorption spectroscopy in ^{28}Si: the importance of inhomogeneous isotope broadening. Phys. Rev. Lett. **90**, 186402 (2003)
17. D. Karaiskaj, M. Thewalt, T. Ruf, M. Cardona, M. Konuma, Photoluminescence studies of isotopically enriched silicon: isotopic effects on the indirect electronic band gap and phonon energies. Solid State Commun. **123**, 87–92 (2002)
18. D. Karaiskaj, M.L.W. Thewalt, T. Ruf, M. Cardona, Photoluminescence studies of isotopically enriched silicon. Phys. Stat. Sol. B **235**, 63–74 (2003)
19. M. Cardona, M. Thewalt, Isotope effects on the optical spectra of semiconductors. Rev. Mod. Phys. **77**, 1173 (2005)
20. M. Steger, A. Yang, D. Karaiskaj, M.L.W. Thewalt, E.E. Haller, J.W. Ager III, M. Cardona, H. Riemann, N.V. Abrosimov, A.V. Gusev, A.D. Bulanov, A.K. Kaliteevskii, O.N. Godisov, P. Becker, H.-J. Pohl, K.M. Itoh, in *Shallow Impurity Absorption Spectroscopy in Isotopically Enriched Silicon*, ed. by W. Jantsch, F. Schaffler. AIP Conference Proceedings **893**, pp. 231–232 (2007)

21. M. Steger, A. Yang, D. Karaiskaj, M. Thewalt, E. Haller, J. Ager III, M. Cardona, H. Riemann, N. Abrosimov, A. Gusev, A. Bulanov, A. Kaliteevskii, O. Godisov, P. Becker, H.-J. Pohl, Shallow impurity absorption spectroscopy in isotopically enriched silicon. Phys. Rev. B **79**, 205210 (2009)

22. M. Steger, A. Yang, M. Thewalt, M. Cardona, H. Riemann, N. Abrosimov, M. Churbanov, A. Gusev, A. Bulanov, I. Kovalev, A. Kaliteevskii, O. Godisov, P. Becker, H.-J. Pohl, J. Ager III, E. Haller, Impurity absorption spectroscopy of the deep double donor sulfur in isotopically enriched silicon. Physica B **401–402**, 600–603 (2007)

23. M. Steger, A. Yang, M. Thewalt, M. Cardona, H. Riemann, N. Abrosimov, M. Churbanov, A. Gusev, A. Bulanov, I. Kovalev, A. Kaliteevskii, O. Godisov, P. Becker, H.-J. Pohl, E. Haller, J. Ager III, High-resolution absorption spectroscopy of the deep impurities S and Se in ^{28}Si revealing the ^{77}Se hyperfine splitting. Phys. Rev. B **80**, 115204 (2009)

24. M. Thewalt, Spectroscopy of excitons and shallow impurities in isotopically enriched silicon—electronic properties beyond the virtual crystal approximation. Solid State Commun. **133**, 715–725 (2005)

25. M.L.W. Thewalt, T.A. Meyer, D. Karaiskaj, M. Cardona, E.E. Haller, J.W. Ager III, H. Riemann, in *Progress in Semiconductor Spectroscopy Using Isotopically Enriched Si*, ed. by J. Menéndez, C.G.V. de Walle. AIP Conference Proceedings **772**, pp. 67–68 (2005)

26. T. Sekiguchi, M. Steger, K. Saeedi, M.L.W. Thewalt, H. Riemann, N.V. Abrosimov, N. Nötzel, Hyperfine structure and nuclear hyperpolarization observed in the bound exciton luminescence of Bi Donors in natural Si. Phys. Rev. Lett. **104**, 137402 (2010)

27. A. Yang, M. Steger, D. Karaiskaj, M. Thewalt, M. Cardona, K. Itoh, H. Riemann, N. Abrosimov, M.F. Churbanov, A. Gusev, A. Bulanov, A. Kaliteevskii, O. Godisov, P. Becker, H.-J. Pohl, J. Ager III, E.E. Haller, Optical detection and ionization of donors in specific electronic and nuclear spin states. Phys. Rev. Lett. **97**, 227401 (2006)

28. M. Thewalt, A. Yang, M. Steger, D. Karaiskaj, M. Cardona, H. Riemann, N. Abrosimov, A. Gusev, A. Bulanov, I. Kovalev, A. Kaliteevskii, O. Godisov, P. Becker, H. Pohl, E. Haller, J. Ager III, K. Itoh, Direct observation of the donor nuclear spin in a near-gap bound exciton transition: ^{31}P in highly enriched ^{28}Si. J. Appl. Phys. **101**, 081724 (2007)

29. A. Yang, M. Steger, T. Sekiguchi, M.L.W. Thewalt, T.D. Ladd, K.M. Itoh, H. Riemann, N.V. Abrosimov, P. Becker, H.-J. Pohl, Simultaneous subsecond hyperpolarization of the nuclear and electron spins of phosphorus in silicon by optical pumping of exciton transitions. Phys. Rev. Lett. **102**, 257401 (2009)

30. A. Yang, M. Steger, M.L.W. Thewalt, T.D. Ladd, K.M. Itoh, E.E. Haller, J.W. Ager III, H. Riemann, N.V. Abrosimov, P. Becker, H.-J. Pohl, in *Nuclear Polarization of Phosphorus Donors in ^{28}Si by Selective Optical Pumping*, ed. by M. Caldas, N. Studart. AIP Conference Proceedings **1199**, pp. 375–376 (2010)

31. M. Steger, T. Sekiguchi, A. Yang, K. Saeedi, M.E. Hayden, M.L.W. Thewalt, K.M. Itoh, H. Riemann, N.V. Abrosimov, P. Becker, H.-J. Pohl, Optically detected NMR of optically hyperpolarized ^{31}P neutral donors in ^{28}Si. J. Appl. Phys. **109**, 102411 (2011)

32. M. Thewalt, M. Steger, A. Yang, M. Cardona, H. Riemann, N.V. Abrosimov, M. Churbanov, A. Gusev, A. Bulanov, I. Kovalev, A. Kaliteevskii, O. Godisov, P. Becker, H.-J. Pohl, Can highly enriched ^{28}Si reveal new things about old defects? Phys. B **401–402**, 587–592 (2007)

33. A. Yang, M. Steger, M. Thewalt, M. Cardona, H. Riemann, N. Abrosimov, M. Churbanov, A. Gusev, A. Bulanov, I. Kovalev, A. Kaliteevskii, O. Godisov, P. Becker, H.-J. Pohl, J. Ager III, E. Haller, High resolution photoluminescence of sulphur- and copper-related isoelectronic bound excitons in highly enriched ^{28}Si. Phys. B **401–402**, 593–596 (2007)

34. M. Steger, A. Yang, N. Stavrias, M. Thewalt, H. Riemann, N. Abrosimov, M. Churbanov, A. Gusev, A. Bulanov, I. Kovalev, A. Kaliteevskii, O. Godisov, P. Becker, H.-J. Pohl, Reduction of the linewidths of deep luminescence centers in ^{28}Si reveals fingerprints of the isotope constituents. Phys. Rev. Lett. **100**, 177402 (2008)

35. M. Steger, A. Yang, M.L.W. Thewalt, H. Riemann, N.V. Abrosimov, P. Becker, H.-J. Pohl, in *High Resolution Photoluminescence of Copper, Silver, Gold and Lithium-Related Isoelectronic Bound Excitons in Highly Enriched ^{28}Si*, ed. by M. Caldas, N. Studart. AIP Conference Proceedings **1199**, pp. 33–34 (2010)

36. M. Steger, A. Yang, T. Sekiguchi, K. Saeedi, M. Thewalt, M. Henry, K. Johnston, H. Riemann, N.V. Abrosimov, M. Churbanov, A. Gusev, A. Bulanov, I. Kaliteevski, O. Godisov, P. Becker, H.-J. Pohl, Isotopic fingerprints of gold-containing luminescence centers in ^{28}Si. Phys. B **404**, 5050–5053 (2009)
37. M. Steger, A. Yang, T. Sekiguchi, K. Saeedi, M.L.W. Thewalt, M.O. Henry, K. Johnston, E. Alves, U. Wahl, H. Riemann, N.V. Abrosimov, M.F. Churbanov, A.V. Gusev, A.K. Kaliteevskii, O.N. Godisov, P. Becker, H.-J. Pohl, Isotopic fingerprints of Pt-containing luminescence centers in highly enriched ^{28}Si. Phys. Rev. B **81**, 235217 (2010)
38. G. Davies, The optical properties of luminescence centres in silicon. Phys. Rep. **176**, 83–188 (1989)
39. K. Shirai, H. Yamaguchi, A. Yanase, H. Katayama-Yoshida, A new structure of Cu complex in Si and its photoluminescence. J. Phys. Cond. Mat. **21**, 064249 (2009)
40. K. Shirai, H. Yamaguchi, J. Ishisada, K. Matsukawa, A. Yanase, S. Emura, in *Cu Complex in Silicon and Its Photoluminescence*, ed. by M. Caldas, N. Studart. AIP Conference Proceedings **1199**, pp. 91–92 (2010)
41. M. Nakamura, S. Murakami, N.J. Kawai, S. Saito, K. Matsukawa, H. Arie, Compositional transformation between Cu centers by annealing in Cu-diffused silicon crystals studied with deep-level transient spectroscopy and photoluminescence. Jpn. J. Appl. Phys. **48**, 082302 (2009)
42. M. Nakamura, S. Murakami, Deep-level transient spectroscopy and photoluminescence studies of formation and depth profiles of copper centers in silicon crystals diffused with dilute copper. Jpn. J. Appl. Phys. **49**, 071302 (2010)
43. S.K. Estreicher, A. Carvalho, The Cu$_{PL}$ Defect and the Cu$_{s1}$Cu$_{i3}$ Complex. Phys. B: Cond. Mat. **407**(15), 2967–2969 (2012)
44. A. Carvalho, D.J. Backlund, S.K. Estreicher, Four-copper complexes in Si and the Cu$_{PL}$ defect: a first-principles study. Phys. Rev. B **84**, 155322 (2011)
45. J.R. Haynes, Experimental proof of the existence of a new electronic complex in silicon. Phys. Rev. Lett. **4**, 361–363 (1960)
46. P.J. Dean, W.F. Flood, G. Kaminsky, Absorption due to bound excitons in silicon. Phys. Rev. **163**, 721–725 (1967)
47. D.F. Nelson, J.D. Cuthbert, P.J. Dean, D.G. Thomas, Auger recombination of excitons bound to neutral donors in gallium phosphide and silicon. Phys. Rev. Lett. **17**, 1262–1265 (1966)
48. W. Schmid, Auger lifetimes for excitons bound to neutral donors and acceptors in Si. Phys. Stat. Sol. B **84**, 529 (1977)
49. P.J. Dean, D.C. Herbert, in *Bound Excitons in Semiconductors, Topics in Current Physics: Excitons*, ed. by K. Cho. vol. 14 (Springer, Berlin, 1979) p. 55
50. M. Thewalt, in *Excitons*, ed. by E.I. Rashba, M.D. Sturge (North Holland, Amsterdam, 1982), pp. 393–458
51. R.E. Dietz, D.G. Thomas, J.J. Hopfield, Mirror absorption and fluorescence in ZnTe. Phys. Rev. Lett. **8**, 391–393 (1962)
52. D.G. Thomas, J.J. Hopfield, C.J. Frosch, Isoelectronic traps due to nitrogen in gallium phosphide. Phys. Rev. Lett. **15**, 857–860 (1965)
53. F.A. Trumbore, M. Gershenzon, D.G. Thomas, Luminescence due to the isoelectronic substitution of bismuth for phosphorus in gallium phosphide. Appl. Phys. Lett. **9**, 4–6 (1966)
54. J.J. Hopfield, D.G. Thomas, R.T. Lynch, Isoelectronic donors and acceptors. Phys. Rev. Lett. **17**, 312–315 (1966)
55. A. Baldereschi, J.J. Hopfield, Binding to isoelectronic impurities in semiconductors. Phys. Rev. Lett. **28**, 171–174 (1972)
56. R. Sauer, J. Weber, Photoluminescence characterisation of deep defects in silicon. Phys. B+C **116**, 195 (1983)
57. E.C. Lightowlers, G. Davies, Spectroscopy of excitons bound to isoelectronic defect complexes in silicon. Solid State Commun. **53**, 1055–1060 (1985)
58. B. Monemar, U. Lindefelt, W.M. Chen, Electronic structure of bound excitons in semiconductors. Phys. B+C **146**, 256–285 (1987)

59. H. Conzelmann, Photoluminescence of transition metal complexes in silicon. Appl. Phys. A **42**, 1–18 (1987)
60. B. Monemar, Electronic structure and bound excitons for defects in semiconductors from optical spectroscopy. Crit. Rev. Solid State Ma. Sci. **15**, 111–151 (1988)
61. J. Weber, W. Schmid, R. Sauer, Excitons bound to an isoelectronic trap in silicon. J. Lumin. **18–19**, 93–96 (1979)
62. J. Weber, W. Schmid, R. Sauer, Localized exciton bound to an isoelectronic trap in silicon. Phys. Rev. B **21**, 2401–2414 (1980)
63. R. Sauer, J. Weber, W. Zulehner, Nitrogen in silicon: towards the identification of the 1.1223-eV (A, B, C) photoluminescence lines. Appl. Phys. Lett. **44**, 440–442 (1984)
64. R.A. Modavis, D.G. Hall, Aluminum-nitrogen isoelectronic trap in silicon. J. Appl. Phys. **67**, 545–547 (1990)
65. J. Wagner, R. Sauer, Acceptorlike excited states of the isoelectronic A, B, C exciton system in silicon. Phys. Rev. B **26**, 3502–3505 (1982)
66. G.S. Mitchard, S.A. Lyon, K.R. Elliott, T.C. McGill, Observation of long lifetime lines in photoluminescence from Si:In. Solid State Commun. **29**, 425–429 (1979)
67. J. Weber, R. Sauer, P. Wagner, Photoluminescence from a thermally induced indium complex in silicon. J. Lumin. **24–25**, 155–158 (1981)
68. M.L.W. Thewalt, U.O. Ziemelis, P.R. Parsons, Enhancement of long lifetime lines in photoluminescence from Si:In. Solid State Commun. **39**, 27–30 (1981)
69. M.L.W. Thewalt, U.O. Ziemelis, R.R. Parsons, Isoelectronic bound excitons in silicon: the role of deep acceptors. Phys. Rev. B **24**, 3655–3658 (1981)
70. T.E. Schlesinger, T.C. McGill, Role of Fe in new luminescence lines in Si:Tl and Si:In. Phys. Rev. B **25**, 7850–7851 (1982)
71. S. Watkins, M. Thewalt, T. Steiner, Isoelectronic bound excitons in In- and Tl-doped Si: a novel semiconductor defect. Phys. Rev. B **29**, 5727–5738 (1984)
72. H. Conzelmann, A. Hangleiter, J. Weber, Thallium-related isoelectronic bound excitons in silicon. A bistable defect at low temperatures. Phys. Stat. Sol. B **133**, 655–668 (1986)
73. T. Schlesinger, R. Hauenstein, R. Feenstra, T. McGill, Isotope shifts for the P, Q, R lines in Indium-Doped silicon. Solid State Commun. **46**, 321–324 (1983)
74. S. Watkins, U. Ziemelis, M. Thewalt, Long lifetime photoluminescence from a deep centre in copper-doped silicon. Solid State Commun. **43**, 687–690 (1982)
75. M.O. Henry, E.C. Lightowlers, N. Killoran, D.J. Dunstan, B.C. Cavenett, Bound exciton recombination in beryllium-doped silicon. J. Phys. C **14**, L255 (1981)
76. R.K. Crouch, J.B. Robertson, T.E. Gilmer, Study of Beryllium and Beryllium-Lithium complexes in single-crystal silicon. Phys. Rev. B **5**, 3111–3119 (1972)
77. G. Davies, A simple model for excitons bound to axial isoelectronic defects in silicon. J. Phys. C **17**, 6331 (1984)
78. M.O. Henry, K.G. McGuigan, M.C. do Carmo, M.H. Nazare, E.C. Lightowlers, A photoluminescence investigation of local mode vibrations of the beryllium pair centre in silicon. J. Phys. Cond. Mat. **2**, 9697 (1990)
79. M.O. Henry, K.A. Moloney, J. Treacy, F.J. Mulligan, E.C. Lighowlers, Uniaxial stress studies of the be pair bound exciton absorption spectrum in silicon. J. Phys. C **17**, 6245 (1984)
80. T. Ishikawa, T. Sekiguchi, K. Yoshizawa, K. Naito, M.L.W. Thewalt, K.M. Itoh, Zeeman photoluminescence spectroscopy of isoelectronic beryllium pairs in silicon. Solid State Commun. **150**, 1827–1830 (2010)
81. N. Killoran, D.J. Dunstan, M.O. Henry, E.C. Lightowlers, B.C. Cavenett, The isoelectronic centre in beryllium-doped silicon. I. Zeeman study. J. Phys. C **15**, 6067 (1982)
82. S. Kim, I.P. Herman, K.L. Moore, D.G. Hall, J. Bevk, Use of hydrostatic pressure to resolve phonon replicalike features in the photoluminescence spectrum of beryllium-doped silicon. Phys. Rev. B **52**, 16309–16312 (1995)
83. S. Kim, I.P. Herman, K.L. Moore, D.G. Hall, J. Bevk, Hydrostatic pressure dependence of isoelectronic bound excitons in beryllium-doped silicon. Phys. Rev. B **53**, 4434–4442 (1996)

84. D. Labrie, T. Timusk, M.L.W. Thewalt, Far-infrared absorption spectrum of be-related bound excitons in silicon. Phys. Rev. Lett. **52**, 81–84 (1984)

85. E. Tarnow, S.B. Zhang, K.J. Chang, D.J. Chadi, Theory of Be-induced defects in Si. Phys. Rev. B **42**, 11252–11260 (1990)

86. M.L.W. Thewalt, S.P. Watkins, U.O. Ziemelis, E.C. Lightowlers, M.O. Henry, Photoluminescence lifetime, absorption and excitation spectroscopy measurements on isoelectronic bound excitons in beryllium-doped silicon. Solid State Commun. **44**, 573–577 (1982)

87. J.H. Svensson, B. Monemar, E. Janzén, Pseudodonor electronic excited states of neutral complex defects in silicon. Phys. Rev. Lett. **65**, 1796–1799 (1990)

88. J. Olajos, M. Kleverman, H. Grimmeiss, High-resolution spectroscopy of silver-doped silicon. Phys. Rev. B **38**, 10633–10640 (1988)

89. F. Bassani, G. Iadonisi, B. Preziosi, Electronic impurity levels in semiconductors. Rep. Prog. Phys. **37**, 1099 (1974)

90. E. Janzen, R. Stedman, G. Grossmann, H. Grimmeiss, High-resolution studies of sulfur- and selenium-related donor centers in silicon. Phys. Rev. B **29**, 1907–1918 (1984)

91. G. Armelles, J. Barrau, M. Brousseau, B. Pajot, C. Naud, Effective mass-like states of the deep acceptor level of Au and Pt in silicon. Solid State Commun. **56**, 303–305 (1985)

92. M. Kleverman, J. Olajos, H.G. Grimmeiss, Observation of $P_{1/2}$ resonant states and Fano resonances of the deep gold acceptor in silicon. Phys. Rev. B **35**, 4093–4094 (1987)

93. M. Kleverman, J. Olajos, H.G. Grimmeiss, Phonon interactions at the deep platinum acceptor in silicon. Phys. Rev. B **37**, 2613–2617 (1988)

94. J.V.W. Morgan, T.N. Morgan, Stress effects on excitons bound to axially symmetric defects in semiconductors. Phys. Rev. B **1**, 739–749 (1970)

95. H.D. Mohring, J. Weber, R. Sauer, Photoluminescence of excitons bound to an isoelectronic trap in silicon associated with boron and iron. Phys. Rev. B **30**, 894–904 (1984)

96. G. Davies, T. Gregorkiewicz, M. Zafar Iqbal, M. Kleverman, E. Lightowlers, N. Vinh, M. Zhu, Optical properties of a silver-related defect in silicon. Phys. Rev. B **67**, 235111 (2003)

97. K. Nishikawa, R. Barrie, Phonon broadening of impurity spectral lines. I. General theory. Can. J. Phys. **41**, 1135 (1963)

98. R. Barrie, K. Nishikawa, Phonon broadening of impurity spectral lines. II. Application to silicon. Can. J. Phys. **41**, 1823 (1963)

99. M. Thewalt, D. Brake, Ultra-high resolution photoluminescence studies of bound excitons and multi-bound exciton complexes in silicon. Mat. Sci. Forum **65–66**, 187–198 (1990)

100. D. Karaiskaj, T. Meyer, M. Thewalt, M. Cardona, Dependence of the ionization energy of shallow donors and acceptors in silicon on the host isotopic mass. Phys. Rev. B **68**, 121201(R) (2003)

101. A.T. Collins, S.C. Lawson, G. Davies, H. Kanda, Indirect energy gap of ^{13}C diamond. Phys. Rev. Lett. **65**, 891–894 (1990)

102. C. Parks, A.K. Ramdas, S. Rodriguez, K.M. Itoh, E.E. Haller, Electronic band structure of isotopically pure germanium: modulated transmission and reflectivity study. Phys. Rev. B **49**, 14244–14250 (1994)

103. D. Karaiskaj, M.L.W. Thewalt, T. Ruf, M. Cardona, M. Konuma, Intrinsic acceptor ground state splitting in silicon: an Isotopic effect. Phys. Rev. Lett. **89**, 016401 (2002)

104. M. Cardona, T.A. Meyer, M.L.W. Thewalt, Temperature dependence of the energy gap of semiconductors in the low-temperature limit. Phys. Rev. Lett. **92**, 196403 (2004)

105. E.S. Johnson, W.D. Compton, J.R. Noonan, B.G. Streetman, Recombination luminescence from electron-irradiated Li-diffused Si. J. Appl. Phys. **44**, 5411–5418 (1973)

106. L. Canham, G. Davies, E.C. Lightowlers, The 1.045 eV vibronic band in silicon doped with lithium. J. Phys. C: Solid State Phys. **13**, L757 (1980)

107. L. Canham, G. Davies, E. Lightowlers, The 1.045 eV vibronic band in irradiated silicon doped with lithium. Inst. Phys. Conf. Ser. **59**, 211–216 (1981)

108. V. Heine, C.H. Henry, Theory of the isotope shift for zero-phonon optical transitions at traps in semiconductors. Phys. Rev. B **11**, 3795–3803 (1975)

109. T.N. Morgan, B. Welber, R.N. Bhargava, Optical properties of Cd-O and Zn-O complexes in GaP. Phys. Rev. **166**, 751–753 (1968)
110. G. Davies, M. Zafar Iqbal, E.C. Lightowlers, Exciton self-trapping at an isoelectronic center in silicon. Phys. Rev. B **50**, 11520–11530 (1994)
111. G. Davies, M. do Carmo, Vibronic coupling in shallow excited states of optical centres in silicon. Inst. Phys. Conf. Ser. **95**, 125–130 (1989)
112. S. Estreicher, D. West, J. Goss, S. Knack, J. Weber, First-principles calculations of pseudolocal vibrational modes: the case of Cu and Cu pairs in Si. Phys. Rev. Lett. **90**, 035504 (2003)
113. S.K. Estreicher, First-principles theory of copper in silicon. Mat. Sci. Semicond. Process. **7**, 101–111 (2004)
114. S. Estreicher, Rich chemistry of copper in crystalline silicon. Phys. Rev. B **60**, 5375–5382 (1999)
115. S. Estreicher, D. West, J. Pruneda, S. Knack, J. Weber, Formation and properties of three copper pairs in silicon. Mat. Res. Soc. Symp. Proc. **719**, 421–426 (2002)
116. S. Estreicher, D. West, M. Sanati, $^*Cu_0$: A metastable configuration of the Cu_s-Cu_i pair in Si. Phys. Rev. B **72**, 121201 (2005)
117. S. Knack, Copper-related defects in silicon. Mat. Sci. Semicond. Process. **7**, 125–141 (2004)
118. E.R. Weber, Transition metals in silicon. Appl. Phys. A **30**, 1–22 (1983)
119. M. Datta, Applications of electrochemical microfabrication: an introduction. IBM J. Res. Dev. **42**, 563–566 (1998)
120. A.A. Istratov, H. Hieslmair, E.R. Weber, Iron contamination in silicon technology. Appl. Phys. A **70**, 489–534 (2000)
121. R.N. Hall, J.H. Racette, Diffusion and solubility of copper in extrinsic and intrinsic Germanium, Silicon, and Gallium Arsenide. J. Appl. Phys. **35**, 379–397 (1964)
122. A.A. Istratov, C. Flink, H. Hieslmair, E.R. Weber, T. Heiser, Intrinsic diffusion coefficient of interstitial Copper in Silicon. Phys. Rev. Lett. **81**, 1243–1246 (1998)
123. M.H. Rashid, *Power Electronics Handbook: Devices, Circuits, and Applications* (Academic Press, Burlington, 2007)
124. K. Graff, H. Pieper, in *Semiconductor Silicon*, ed. by H.R. Ruff, R.S. Kriegler, Y. Takeishi (Electrochemical Society, Pennington, 1981)
125. H. Reiss, C.S. Fuller, F.J. Morin, Bell Syst. Tech. J. **35**, 535 (1956)
126. J.D. Struthers, Solubility and diffusivity of Gold, Iron, and Copper in Silicon. J. Appl. Phys. **27**, 1560–1560 (1956)
127. R. Keller, M. Deicher, W. Pfeiffer, H. Skudlik, D. Steiner, T. Wichert, Copper in silicon. Phys. Rev. Lett. **65**, 2023–2026 (1990)
128. A. Mesli, T. Heiser, Defect reactions in copper-diffused and quenched p-type silicon. Phys. Rev. B **45**, 11632–11641 (1992)
129. T. Heiser, A. Mesli, Determination of the copper diffusion coefficient in silicon from transient ion-drift. Appl. Phys. A **57**, 325–328 (1993)
130. A. Mesli, T. Heiser, E. Mulheim, Copper diffusivity in silicon: a re-examination. Mat. Sci. Eng. B **25**, 141–146 (1994)
131. P. Wagner, H. Hage, H. Prigge, T. Prescha, J. Weber, in *Semiconductor Silicon*, ed. by H.R. Huff, K.G. Barraclough, J.-I. Chikawa (Electrochemical Society, Pennington, 1990), p. 675
132. D.E. Woon, D.S. Marynick, S.K. Estreicher, Titanium and copper in Si: barriers for diffusion and interactions with hydrogen. Phys. Rev. B **45**, 13383–13389 (1992)
133. H. Prigge, P. Gerlach, P.O. Hahn, A. Schnegg, H. Jacob, Acceptor compensation in silicon induced by chemomechanical polishing. J. Electrochem. Soc. **138**, 1385–1389 (1991)

Chapter 2
History of the Observed Centres in Silicon

Many of the IBE centres studied in this thesis have been discussed in the past. While none of those prior publications could ultimately determine the actual constituents of these complexes, a number of important properties have been determined, over the years, through various different experiments. Many of these characteristics are introduced in this chapter for the "Cu-pair" (Cu_4), the "*Cu-pair" (Cu_3Ag), the "single Ag" (Ag_4), the "Au-Fe pair" (Cu_3Au), the "Fe-B pair" (Cu_4Au), and the Cu_4Pt and Cu_3Pt complexes. The related S_A and S_B centres as well as the Li_4-vacancy centres, are introduced as well.

2.1 The 1014 meV "Cu-pair": The Cu_4 Centre

Among all TM-containing IBE centres, the $\sim$1014 meV Cu_4 complex, the NP transition of which is often labelled Cu_0^0, has been studied the most in recent literature, both experimentally and theoretically, and the complex can be considered the prototype of the entire family of centres. In fact, it is difficult not to observe this PL centre after rapidly quenching Si from above 700 °C to room temperature. This is due to the high diffusivity and solubility of Cu at elevated temperature, together with the difficulty of eliminating Cu as a contaminant in all but the most rigorously clean environments. The centre was first observed by Minaev et al. [1] at 1015 meV (various authors refer to the centre as the 1014, 1015 meV, or more accurately, the 1014.7 meV centre) in heat treated and subsequently quenched n-type Si. In view of the technique used to create this centre, the authors called it a "thermal defect". The low temperature, low resolution spectra clearly show the intense NP line, together with up to 10 lower energy pLVM replicas. At a temperature above 10 K, higher energy components were observed. Using uniaxial stress spectroscopy, pLVM replicas and the NP transition split in the same way, indicating totally symmetric vibrations. Minaev et al. concluded that this luminescence is due to the recombination of an exciton bound to a trigonal, neutral centre (an IBE), with $\langle 111 \rangle$ axial symmetry. It was also

M. Steger, *Transition-Metal Defects in Silicon*, Springer Theses,
DOI: 10.1007/978-3-642-35079-5_2, © Springer-Verlag Berlin Heidelberg 2013

found that this defect anneals out at less than 300 °C, with an activation energy of 0.6 eV and has an estimated formation energy of 2.6 eV. It was concluded that the defect is likely due to the presence of TM impurities in the Si sample, with Fe and Cr as possible candidates. The centre was then briefly mentioned by Weber et al. [2] as a PL centre at $\sim$1016 meV, which was found to be especially strong in Cu-doped samples.

Shortly after these initial reports, Weber et al. [3] published a comprehensive study of the $\sim$1014 meV Cu PL centre. The authors systematically assigned labels to NP, Stokes and anti-Stokes vibrational replicas, as well as to electronically excited states of the NP transition detected at higher temperatures. The characteristic pLVM structure formed by the Stokes vibrational replicas was described in detail in this study. The investigated samples were prepared by intentionally evaporating Cu onto them, before subjecting them to diffusion steps in a furnace. No increase in PL intensity was observed after more than 5 min of diffusion time for a sample thickness of $\sim$1 mm. A high cooling rate was necessary for strong PL emission, as already noted by Minaev et al. [1]. PL was observed for samples diffused above 700 °C with an increasing intensity up to 1100 °C. No difference in the experimental results was found with different n or p-type doping levels of P, B, or Al in the starting Si material for concentrations of either of the dopants in the range between 10^{12} and 10^{17} cm^{-3}. The intensity of the Cu spectrum decreased for concentrations above 10^{17} cm^{-3}, but no details were given [3]. It was also confirmed that the centre anneals out at temperatures above 150 °C. For the first time, isotopic Cu was used to show clearly the role that Cu plays in the 1014 meV PL line. A total isotope shift of 0.06 meV was observed for the NP line, which compares very well to the experimental result obtained in this work (a total shift of 0.06 meV corresponds to 7.5 μeV per amu as shown in Table 4.2), and proved that Cu is indeed part of this PL centre [3]. An isotope shift was also observed in the first order pLVM replica. With a ratio of 1.010 of the lighter over the heavier mode energy, this value is comparable to the shift determined in this work (see Table 4.3). In an attempt to determine the number of Cu atoms needed to form the 1014 meV PL centre, experiments were done to find a relationship between the PL intensity I and the Cu concentration N_{Cu} in the sample. To determine the Cu concentration, the relationship between diffusion temperature and Cu concentration as determined by Struthers [4] was used (see Sect. 1.6.3). These experiments were conducted at the high temperature Cu solubility limit. After a quadratic dependence was found ($I \propto N_{Cu}^2$), conclusions were made based on the law of mass action that the centre was composed of two Cu atoms [3]. Thus the complex was often referred to as the "Cu-pair" centre until it was demonstrated to be a Cu$_4$ complex by isotopic fingerprint measurements in ^{28}Si (see below).

Weber et al. [3] also observed the weak Γ_3 excited state, as a shoulder $\sim$0.15 meV above the main Γ_4 peak. The higher energy NP lines (Γ_3 and Γ_5 at 1016 meV) were identified as thermally populated split ground states of the trap. The localization energy of the BE of Cu$_4$ was found to be 140.3 meV with respect to the free exciton (FE) or 155 meV with respect to the band gap energy (the FE binding energy is 14.7 meV). As determined in a thermalization experiment, the donor electron is bound by 32 meV, whereas the hole is bound by 123 meV in the short range potential of the isoelectronic trap. This finding agrees reasonably with a hole trap at $E_v + 0.1$ eV, which was found in a deep-level transient spectroscopy (DLTS) experiment involving Cu-doped Si [5] at about the same time. The authors concluded that the centre is a donor-like IBE. This argument is further supported by the observed high quantum efficiency of the centre. Through uniaxial stress and Zeeman spectroscopy experiments, and assuming a Cu-pair, the authors verified that the IBE centre is in a $\langle 111 \rangle$ trigonal configuration with at least one Cu on an interstitial site (Cu_sCu_i). The centre reorients itself along the $\langle 111 \rangle$ axis under stress, which again suggests the involvement of a highly mobile Cu_i.

Almost simultaneously, Watkins et al. [6] found that the 1014 meV PL centre is present at low Cu concentration, but has a high PL efficiency. The authors focused on the details of the pLVM replicas, and on the temperature dependence of the PL lifetime. In addition to the ~ 7 meV mode, and the various overtone and combination modes, distinct pLVM were observed with energies of 16.4 and 25.1 meV. The measured long PL decay times of the centre were consistent with an IBE model. Decay times of up to 480 μs were measured at 1.3 K and up to 670 μs at 4.2 K. The measured temperature dependence of the lifetime was later explained by Sauer et al. [7] to fit the level scheme for the split ground state suggested in reference [3], where the excited Γ_3 state has a very long lifetime, which explains the longer measured lifetimes at higher temperature.

Davies [8] reviewed a number of PL centres in Si, among them the 1014 meV PL complex. This publication noted that the structure of the centre was not known, even though it was thought to be a "Cu-pair" [3]. It was unclear though, whether all of the Cu dissolved in the Si sample actually forms the 1014 meV PL centre, and, thus, a connection between Cu concentration and PL intensity could not be made [8].

The connection between this PL centre and the $E_v + 0.1$ eV hole trap observed by DLTS was further supported by Brotherton et al. [9], and firmly established by Erzgräber et al. [10], who found a linear relationship between the PL saturation intensity of the 1014 meV centre and the concentration of the $E_v + 0.1$ eV DLTS centre. The 1014 meV centre is one of the few examples of a firm connection between a deep defect seen in PL and one seen in a DLTS experiment.

Nazaré et al. [11] have investigated the excited states of the 1014 meV centre at 16 K under uniaxial stress and confirmed that the centre is an IBE with trigonal C_{3v} symmetry. In a different interpretation of the system than the one given by Weber et al. [3], it was found that a built-in axial strain splits the ground state of the centre into the Γ_5 (1014 meV, $\Gamma_5 = \Gamma_1 + \Gamma_3$) state together with higher lying Γ_1 (1016 meV) and Γ_3 (1024 meV) states. Both interpretations are consistent with a donor-like IBE model.

By this time, many authors were referring to the 1014 meV centre as the Cu-pair centre [12–17]. Istratov et al. [12] found the binding energy of the 1014 meV Cu PL centre was 1.02 eV in a DLTS experiment. With the assumptions that interstitial Cu_i is singly positively charged and substitutional Cu_s is doubly negatively charged (fitting the measured binding energy), the only solution that fits a donor-acceptor pair model for the Cu-pair is that the centre is a single acceptor ($Cu_s^{2-}Cu_i^+$), contradicting the commonly accepted model of the centre being donor-like. The authors also found that the pair formation is limited by the availability of Cu_s. Shortly thereafter, Istratov et al. [13] revised these results, as they found that the two Cu atoms in the Cu-pair were not exclusively bound Coulombically, thus invalidating the previous assumption of a donor-acceptor pair model. This new result is based on the fact that in a purely ionic model, the binding energy between Cu_s^- and Cu_i^+ is only 0.52 eV, half of the experimentally determined value of 1.02 eV [12, 13]. In a double deep-level transient spectroscopy (DDLTS) experiment, the possibility of the Cu-pair being an acceptor was now excluded, therefore the centre has to be a donor-like IBE [3]. However, the only possibility to have a dissociation energy of 1.02 eV for an ionic Cu pair is if it was an acceptor [13]. Thus the centre was argued to have a strong covalent contribution to its binding energy.

In contrast to the Cu-pair model, Nakamura et al. [18] started arguing for a single Cu model. Their initial argument was based on the measured relationship between the PL intensity of the centre and the Cu concentration (which was kept below the solubility limit) in the crystal. The Cu concentration in the crystal was determined through a measurement of the surface concentration together with a constant-concentration depth profile at the used diffusion temperature. The linear relationship that was determined contradicts the results of Weber et al. [3], which were obtained at much higher concentrations. No difference was found for n or p-type samples for B and P doping concentrations between 5×10^{14} and 2×10^{15} [18]. Nakamura et al. suggested a bond-centred (BC) model for a single Cu atom, which satisfies the trigonal symmetry requirement, but the authors realized possible interference with the rapid rearrangement under stress, which had been suggested before [3]. In a following publication [19], this problem was clarified by measuring a small dissociation energy for the centre that the authors thought would allow for the reorientation. In a later publication, Nakamura et al. [20] suggest that a DLTS centre at $E_c - 0.15$ eV is a precursor to the 1014 meV Cu PL centre (which corresponds to the $E_v + 0.1$ eV DLTS centre). The authors also found that the dissociation energy of the 1014 meV centre was only 0.63 eV in a PL experiment, as compared to the value of 1.02 eV determined by Istratov et al. [13] in a DLTS experiment. The smaller value was thought to be erroneous [15, 17], as it was measured by PL, and the long exciton diffusion length would contribute a signal from deeper regions of the bulk; whereas Istratov et al. measured the dissociation energy with DLTS, where only a near-surface region is probed, thereby increasing the likelihood that only material is probed in which the Cu_i had a high chance to out-diffuse. The out-diffusion process is likely slower in the deeper bulk probed by PL, and thus PL would only measure the effective dissociation. Nakamura et al. [21] confirmed this suggestion, and explained the discrepancy between the different values of the measured dissociation energy in a depth-dependent DLTS experiment. The formation energy of the Cu_4 centre of 0.57 eV, as determined by Nakamura et al. [20] via DLTS, was found to be slightly less than the measured DLTS dissociation energy, indicating that the formation process is different than the dissociation process.

Knack et al. [15] then undertook a number of experiments to achieve a better understanding of the formation mechanism of the 1014 meV Cu PL centre. A sample with Cu_s incorporated during the floating-zone growth process was used. After initially observing the $E_v + 0.1\,\text{eV}$ DLTS peak, only a DLTS signal due to Cu$_s$ and Cu-H complexes was observed after a 30 min annealing process at 250 °C. Cu$_s$ has no corresponding luminescence signal. A Cu-containing etch solution restored the $E_v + 0.1\,\text{eV}$ DLTS peak. The authors concluded that Cu$_i$ diffuses out of the sample during the annealing step, whereas it diffuses back into the sample during the room temperature etch process and then binds to the still available substitutional Cu$_s$, which forms only at higher temperatures (>700 °C). Thus, the concentration of the complex is $[\text{Cu}_s\text{Cu}_i] = [\text{Cu}_s] = k[\text{Cu}_{tot}]$, where the solubility of the substitutional Cu$_s$ determines the saturation of the pair formation, not the much higher total solubility, which is dominated by interstitial Cu$_i$. The authors [15] argued that this model can comply with the linear relationship between PL intensity and Cu concentration for a Cu content in the sample below the solubility limit, as observed by Nakamura et al. [18]. Regarding the quadratic dependence of the PL signal at higher concentrations at the Cu solubility limit determined by Weber et al. [3], Knack et al. [15] show that literature values for $[\text{Cu}_i]$, as well as the Si vacancy concentration and the formation enthalpy of Cu$_s$, lead to a near quadratic dependence of the PL intensity on the Cu concentration, which is controlled by the Cu$_s$ concentration, not the total Cu concentration.

The similarity between the 1014 and the 944 meV Cu related PL centres (see Sect. 2.2) was noted by Estreicher et al. [22], and their formation and vibrational properties were discussed. For the 1014 meV centre, the authors ruled out Cu-pairs of either two interstitial or two Cu$_s$, and also determined that a BC site, as was suggested by Nakamura et al. [18], is not energetically favorable for Cu in Si. It was found from *ab-initio* calculations, that the energy gained by a Cu atom taking the place of a vacancy in Si is between 2.71 and 2.78 eV. On this substitutional site, the Cu-Si bond length to the four Si neighbours was determined to be 2.236 Å, within $\sim$0.1 Å of Si-Si bonds. Trapping of a Cu$_i$ by a Cu$_s$ would then result in an additional gain of 0.75 eV. Potentially, this Cu$_s$Cu$_i$ pair can react with another vacancy and form a Cu$_s$Cu$_s$ pair, possibly with a Si in between (Cu$_s$-Si-Cu$_s$), resulting in a further energy gain of more than 2 eV. The authors calculated the energy of a vibrational mode for these models, where both Cu atoms are moving together along the trigonal axis, preserving their bond length. The result is close to the experimentally measured pLVM energies of the 1014 and the 944 meV Cu centres. These results strongly suggest that the 1014 meV is a Cu$_s$Cu$_i$ pair (whereas the 944 meV centre was thought to be a Cu$_s$Cu$_s$ pair, see Sect. 2.2).

The association of the 1014 meV PL centre with a Cu$_s$Cu$_i$ pair was supported later by Estreicher et al. [23] with *ab initio* calculations involving the centre's vibrational modes. One pLVM mode was found for a Cu$_s$Cu$_i$ that has appropriate symmetry to couple to the electronic transition and has a frequency close to the experimentally observed vibrational replica of the 1014 meV centre. The analysis also showed that neither Cu$_s$Cu$_s$, nor Cu$_s$-Si-Cu$_s$, nor Cu$_s$ alone could produce the observed vibrational mode structure. These findings were confirmed in further reviews [17, 24].

Nakamura et al. [25] claimed that the substitutional model was unlikely due to the high concentration of Cu centres detected in DLTS and the small number of available Cu$_s$ in Si at the diffusion temperatures typically used to produce the 1014 meV

PL centre. An experiment compared the Cu PL intensities in Si samples with, and without, vacancies, and found them to be very similar. Thus it was concluded that the presence of vacancies does not increase the formation of the Cu PL centre, even though it should facilitate the formation of substitutional centres, hence making Cu_s an unlikely part of the Cu PL centre. As a consequence, Nakamura et al. [26] kept arguing in favour of a single bond-centred Cu_{BC} model. Their argument continued to be based on the linear relationship of the Cu PL intensity and the DLTS concentration, where the maximum concentration of the centre was thought to be larger than the Cu_s solubility. They concluded that a single Cu_{BC} is more appropriate than the proposed pair models [24], in contrast to calculations that show Cu_{BC} to be unstable [22, 23].

Thewalt et al. [27] then discovered that the 1014 meV Cu centre consists not only of one or two Cu atoms, but a cluster of four Cu atoms. The experimental evidence for this surprising discovery was found through high resolution PL spectroscopy of isotopically enriched ^{28}Si, which had been shown before to exhibit drastically reduced PL linewidths of shallow impurities due to the removal of inhomogeneous isotope broadening [28–32]. With the increased spectroscopic resolution it was now possible to observe the PL lines that originate from different combinations of the stable Cu isotopes in the impurity cluster [27]. Five PL lines could be observed for the Γ_3 representation, leading to the conclusion that they are due to the five possible arrangements of four Cu atoms of the naturally occurring isotopes ^{63}Cu and ^{65}Cu. This conclusion was tested by diffusing monoisotopic Cu and non-natural isotopic mixtures as well as by comparing the relative areas of the peaks in the ^{nat}Cu spectrum to the natural abundances of the Cu isotopes. The controversy about the 1014 meV Cu centre was settled by these unambiguous experimental results, which are discussed in detail in Sect. 4.1.1.

Nevertheless, in a further publication, Nakamura et al. [33] assume a BC model of a single Cu atom forming the 1014 meV centre. Contrary to Weber et al. [3], the authors found that the concentration of PL centres is not a simple increasing function of diffusion temperature in Cu saturated samples. A sharp drop-off in concentration at higher temperatures was explained as a result of increased precipitation and out-diffusion. Again, the authors measured a higher concentration of PL centres than what was thought to be the saturation concentration of Cu_s, making a single Cu_{BC} atom their favoured configuration [33]. In light of the ^{28}Si high resolution spectroscopy results [27], this model was revised finally and Nakamura et al. [34] suggested a model for the Cu_4 centre, in which one BC Cu is surrounded by three equatorially bound Cu_i. The authors concede that this model is in contradiction to calculations regarding the stability of the BC site by Estreicher et al. [23]. In a further publication [35], the authors argue again in favour of $Cu_{BC}Cu_{3i}$ model, based on the argument that the BC site would be stable at higher temperatures and "freeze in" at lower temperature. They argue that centres involving Cu_s are not a realistic model for the 1014 meV line, as there would not be a sufficient number of vacancies available to form Cu_s at the observed concentrations, and there would be a large discrepancy between the Cu_sCu_{3i} binding energy [36] and Nakamura's observed dissociation energy [20]. The BC model was reiterated in a further publication [21], although none of these papers contain any detailed calculations supporting the stability or formation mechanism of the proposed centre.

After it was found that the 1014 meV PL centre is a Cu_4 centre, another model for the structure of the centre was proposed. Shirai et al. [36, 37] suggested a model where three Cu_i are grouped around a Cu_s in C_{3v} symmetry. This arrangement has the strongest binding energy of a number of calculated combinations, but its formation energy has not been determined, therefore different configurations cannot be excluded. The authors also attempt to explain the observed splitting in the Γ_4 representation, by taking vibronic coupling to non-totally symmetric vibrations into account, but a complete explanation of these splittings could not be given. Possible pLVM of the Cu_sCu_{3i} complex have also been studied [37]. Very recently Estreicher et al. [38] and Carvalho et al. [39] explained the formation pathway of the Cu_4 defect, in terms of the capture of the positively charged Cu_i by Cu_s, Cu_sCu_i, and Cu_sCu_{2i} which were all found to be negatively charged, if the Fermi-level was near mid-gap. The Cu_sCu_{3i} centre, on the other hand, has its acceptor level almost resonant with the conduction band, and would be uncharged in intrinsic Si. The model thus explains why no Cu_sCu_i, Cu_sCu_{2i} or Cu_sCu_{4i} centres are observed. In this regard it is important to observe that the samples used here, and to the best of our knowledge in all previous studies, were lightly doped, with Fermi-levels near mid-gap, with the single exception of the brief comment by Weber et al. [3], that the intensity of the PL of the 1014 meV centre decreases in intensity for doping levels above $10^{17}\,cm^{-3}$.

2.2 The 944 meV "Perturbed Cu-pair": The Cu_3Ag Centre

The $\sim$944 meV Cu_3Ag complex, formerly labelled $^\star Cu$, was long thought to be a Cu-pair similar to the Cu_0^0 centre. This centre was likely first observed by Minaev et al. [1], who reported PL lines at 950 and 875 meV in thermally quenched Si, in their paper which first reported the very similar 1014 meV Cu_4 centre. These energies are very close to those of Cu_3Ag and Cu_2Ag_2 (see Sect. 4.1.3), therefore, it is assumed that they are the same centres. The 944 meV centre was described in detail by McGuigan et al. [40] in Si lightly doped with Cu, and found to have a pLVM structure very similar to that observed for the 1014 meV Cu_4 centre. Extensive experiments made the authors believe that this IBE PL line could only be observed in Si that was lightly doped with Cu. Due to the similarity of the two centres, it was thought that this centre is a precursor to Cu_4 at lower Cu concentrations, before a high enough Cu concentration is present in the sample to form Cu_4. It was also found that the centre anneals out at temperatures lower than those necessary for the Cu_4 centre. The involvement of elements other than Cu was not ruled out.

In a later report, McGuigan et al. [41] investigated the defect in uniaxial stress and Zeeman experiments. The authors found that their uniaxial stress and preliminary Zeeman data was consistent with a high symmetry T_d defect centre. Because they could observe the centre only at low Cu concentrations, they suggested that the $\sim$944 meV PL centre was a singly positively charged Cu_i, at a tetrahedral interstitial site with T_d symmetry. In this state, Cu has a full $3d$ shell and creates a single level. However, it was concluded that more data was needed to fully identify this centre. The proposed interstitial Cu_i character of the centre has later been the subject of controversy [17, 23, 42]. As Cu_i is not stable at room temperature, it would instead diffuse out to the surface of the Si sample. In a later publication [43], members of the group of reference [41] put the measured T_d symmetry in doubt as well, because the results were obtained only for low uniaxial stresses.

Later, Knack et al. [15] suggested a connection between the 944 meV *Cu PL centre and a DLTS centre at $E_v + 0.185\,eV$. While going through different treatment and annealing steps, these two centres had always been observed simultaneously in their samples. The authors drew no conclusion about the microscopic origin of the centre.

On the basis of formation dynamics, energetics, symmetry, and low frequency pLVM, Estreicher et al. [22] initially suggested that *Cu is a Cu_sCu_s pair (see Sect. 2.1), which was later ruled out [23] because it was not possible to find matching pLVM modes for the Cu_sCu_s pair that would couple to PL and thus be visible in experiments to generate the characteristic pLVM replica spectrum [23, 24]. In his review, Knack [17] noted that the proposed substitutional pair would have trigonal C_{3v} symmetry, as opposed to the observed tetrahedral T_d symmetry [41], as T_d only allows for a single Cu_i or Cu_s, both of which can be ruled out for several reasons [42].

The most recent modeling by Estreicher et al. [42] revised the previous findings and identified *Cu as a Cu_sCu_i centre in a different configuration than that of the 1014 meV Cu_4 centre. This new configuration is Cu_s-Si-Cu_i as opposed to Si-Cu_s-Cu_i, which was suggested for the 1014 meV centre [23, 24]. Both configurations have C_{3v} symmetry. A pLVM that couples to PL was found close to the measured energy for this new configuration. In this mode, as is the case for the 1014 meV mode, three atoms (Si $+$ 2 Cu) move along the trigonal axis, which explains the strong similarity in the optical spectra of *Cu and Cu_4. It was suggested that further experiments be undertaken to examine the centre's symmetry and formation conditions.

A lot of the efforts to explain *Cu have been hampered by what were thought to be the conditions needed to produce strong PL from the centre. There were uncertainties in the necessary Cu concentration to produce the centre, its appearance in conjunction with dislocations, special annealing and quenching procedures, or the assumption that *Cu would be a precursor to the 1014 meV PL centre. These speculations were put to an end when the inconsistent formation conditions of the *Cu centre were shown to be connected to the presence of Ag in the centre [44, 45] as detailed in Sect. 4.1.2. Hence, in the past, *Cu was always observed under conditions with sufficient accidental Ag contamination. Through the isotopic fingerprint characterisation method in ^{28}Si, Yang et al. [44] unambiguously identified the 944 meV PL centre as Cu_3Ag.

The photoionization of the $\sim$944 meV centre was studied using time-resolved free-electron laser spectroscopy by Vinh et al. [43], in samples where the $\sim$944 meV centre PL signal was much stronger than the normally dominant $\sim$1014 meV centre. It is interesting to note that the authors also observed the 778 meV Ag-related PL system in this sample, and had been preparing Ag-doped samples for similar studies [46]. Given that the isotopic fingerprint result proves that the $\sim$944 meV *Cu centre is, in fact, a Cu_3Ag complex, the surprising strength of the $\sim$944 meV system in the sample studied by Vinh et al. [43] is readily understood. This centre also provides a good example of the difficulty in avoiding cross-contamination when preparing TM-doped samples.

2.3   The 778 meV "Single Ag": The Ag_4 Centre

Ag is an extensively studied contaminant in Si as it is electrically active. The $\sim$778 meV Ag PL system was first observed by Olajos et al. [47] in absorption experiments. The centre was linked to Ag, and the authors suggested that the symmetry of the centre is C_{2v} or lower. The authors also suggested that the centre is an isolated, single substitutional Ag donor-like IBE centre and hence a link to the DLTS Ag donor level $Ag(D)$ at $E_v + 0.34$ eV [48] was established.

Nazaré et al. [49] reported on IBE centres in PL spectra of Si samples diffused and implanted with Ag at 778.9 and 777.3 meV, and found that both lines are thermalizing excited defect states, decaying to the same ground state. Upon closer examination, the reported PL spectrum bears some similarity to Pt related spectra [50], especially when considering the reported pLVM energy of 9.5 meV and, hence, an accidental Pt contamination of the samples used in reference [49] cannot be excluded.

Some of the earlier results were confirmed by Son et al. [51] in PL and absorption experiments. The authors proposed a donor-like IBE model, and also studied the pLVM replica spectrum in which the lowest energy mode was found at 6 meV. Three NP lines originating from the Ag-centre ground state manifold, labelled A, B, and C were observed, and a table listing pLVM and EMA excited state energies was published. These results are also summarized in reference [52]. Son et al. [53, 54] stated that even though Ag-related centres were identified in electron paramagnetic resonance (EPR), no correlation between magnetic resonance and optical spectra could be established.

At about the same time, Iqbal et al. [55], in a study of the IBE ground state splittings, showed the first evidence of the 778.7 meV F_0 forbidden transition $\sim$0.2 meV below the A transition at low temperature. With temperature-dependent intensity measurements, it was shown that the F_0, A, B, C PL lines thermalize and originate from one split ground state of the IBE, and that the dominant pLVM replicas originate from the newly discovered F_0 level. A trigonal symmetry of the defect centre was suggested on the basis of uniaxial stress experiments.

In an experiment where different Ag isotopes were used for the first time to form the 778 meV centre, Vinh et al. [56] reported an Ag-isotope shift for the A, B, C components. A total isotope shift of 0.034 meV was observed between ^{107}Ag and ^{109}Ag, corresponding reasonably well to a new result of 0.025 meV obtained in ^{28}Si ([45] and Sect. 4.1.4). The authors also concluded, on the basis of lifetime measurements that the centre is an IBE centre. A lifetime of up to 250 μs was determined at low temperature, which is long when compared to the lifetimes of regular donor or acceptor BE in Si that typically range on the ns scale. At higher temperature (25 K), additional

PL bands D and E were identified. For this analysis, it was assumed that the centre contained a single Ag atom.

Davies et al. [57] studied the Ag centre in detail. Zeeman spectra were observed to be near isotropic, thus no information could be extracted about the point group of the centre. This result is expected for a tightly bound hole in C_{3v} symmetry, because its orbital angular momentum is quenched and the remaining spin angular momentum responds isotropically to the magnetic field [58]. Uniaxial stress experiments confirmed the EMA character of the electron and the centre was characterised as a donor-like IBE with trigonal C_{3v} symmetry. Temperature-dependent intensity measurements associated the pLVM replicas with the forbidden F_0 state just below the A line, redefining the phonon energy slightly. Due to the energetic closeness of A and F_0, their pLVM replicas could overlap, but it was found that the pLVM replica luminescence is dominated by the F_0 state. The measured isotope shifts of the pLVM and calculations of the effective mass of the centre responsible for the observed pLVM isotope shifts made the authors conclude that the centre consisted of a single Ag atom.

Only experiments with isotopically enriched ^{28}Sicould finally show the nature of the 778 meV PL centre [45]. It was found to consist of four Ag atoms and a detailed analysis is presented in Sect. 4.1.4.

2.4 The 735 meV "Fe, or Au-Fe pair": The Cu_3Au Centre

In 1973, Otha [59] published extensive results on infrared absorption spectroscopy of Au in Si, suggesting the occurrence of single Au centres causing broad absorption bands and Au pair centres causing sharper absorption lines. It was proposed that the Au pair would likely occupy second nearest neighbour sites in the Si lattice.

After these first spectroscopic experiments involving Au defect centres in Si, the $\sim$735 meV centre (now known to be Cu_3Au) was discussed thoroughly in the literature. The first occurrence appeared in a publication on Fe in Si [2], as the centre was believed to be due to Fe, and the Au and Cu content in the sample was likely due to contamination. The 735 meV PL line's characteristic pLVM replicas were observed. Two thermalizing NP lines were detected and a BE model was proposed. The symmetry was determined to be trigonal using Zeeman spectroscopy.

The participation of Fe in the 735 meV centre was put in doubt by Schlesinger et al. [60], because no Fe isotope shift could be observed in their experiments using ^{54}Fe and ^{56}Fe diffused Si samples.

Do Carmo et al. [61] found the 735 meV PL centre was trigonal on the basis of uniaxial stress tests. The centre was interpreted as an IBE centre with a tightly bound hole and a shallow electron, making it a donor-like IBE. Similar to Ag_4 [58], the hole angular momentum is quenched in the low symmetry environment of the optical centre and, as a result of the axial strain, thereby leading to singlet-triplet splitting of the bound exciton. Thermalization of the resulting three electronic states was observed. No attempts to identify the constituents were undertaken.

Singh et al. [52] claimed that the occurrence of the 735 meV PL peak in their Au doped samples was a result of the Fe contamination of even the purest Au source and that the centre is unlikely to contain Au, which had been shown in Fe diffused samples before. Earlier results were interpreted [61] and explained with additional data. Besides the already observed three electronic states that were attributed to the $1s(A_1)$ split states, the authors found two higher excited states and predicted the presence of a third. With this information, the hole binding energy was estimated and found to be close to the level observed for Fe_i donors in Si. Therefore the authors suggested that the defect contains an interstitial Fe atom and possibly other unknown constituents, with an overall trigonal symmetry. The electronic structure would be dominated by Fe and the centre would fit into the donor-like IBE model.

In the first publication where the decay of ion-implanted, radioactive Au isotopes in Si was studied, Henry et al. [62] found that the 735 meV centre decays according to the half life time of the implanted radioisotope ^{195}Au. It was concluded, therefore, that Au was involved in the formation of this centre and it was relabeled the Au-Fe centre [50, 63].

PL spectroscopy of Cu and Au-diffused ^{28}Si samples revealed the presence of Au and 3 Cu atoms in this centre, while no fingerprint of Fe was found [45]. Detailed experimental results for this Cu_3Au centre are presented in Sect. 4.2.1.

2.5   The 1066 meV "FeB pair": The Cu_4Au Centre

Initially, the $\sim$1066 meV Cu_4Au centre was thought to be an FeB pair centre [2], based on observing it in Fe diffused p-type samples. Sauer et al. [64] observed two thermalizing lines, which were both split into a thermalizing triplet under a magnetic field. The two components were interpreted as one forbidden and one allowed transition from an IBE state to the ground state. The centre was found to have trigonal symmetry and pLVM replicas were detected for both PL lines. The authors also confirmed earlier results that linked this 1066 meV PL centre to a DLTS centre at $E_v + 100$ meV, which was ascribed to FeB pairs [65, 66].

Schlesinger et al. [67] have done experiments to determine a possible Fe isotope shift of the 1066 meV PL line and its phonon vibrational modes. Samples diffused with ^{54}Fe and ^{56}Fe did not show any detectable isotope shift in NP and pLVM replica PL lines. This result was not expected for an Fe containing defect, especially for its pLVM, which were described as due to Fe motion [64].

Mohring et al. [68] did extensive research on the 1066 meV PL centre. Three electronic states and their pLVM replicas were observed. Zeeman spectra showed no significant anisotropy, limiting the possible symmetries to cubic T_d, as all lower symmetries would show some anisotropy. It was found that the centre decays at annealing temperatures of more than 80 °C and the centre was not observed in samples with low B concentration, while its intensity was found to be enhanced with higher B doping levels. Hence, Mohring et al. [68] assumed that B is part of the centre. Diffusion experiments also suggested the participation of Fe in the formation of the centre. The determined dissociation energy of 0.9 eV was close to values previously determined for FeB pairs in other experiments. The produced PL samples were also used for DLTS studies, and only a limited correlation was found with the $E_v + 100$ meV centre. Mohring et al. [68] also pointed out points of conflict with a possible FeB assignment of the 1066 meV PL centre. For one, Fe diffuses interstitially, so a pair would likely be B_s-Fe_i and thus aligned along a $\langle 111 \rangle$ axis with axial symmetry, which would show some anisotropy in Zeeman spectra, which was not observed. The PL intensity was relatively low for the expected overall number of centres, taking the amount of available B in the sample into account. The authors concluded that the composition of the centre could not be identified, but their experimental data was inconsistent with an FeB assignment.

After experiments with B and Fe doped Si, Conzelmann et al. [69] concluded that it seems necessary to have Fe in the sample to produce the 1066 meV PL centre ("FeB"), but admitted that his results were "controversial" in light of some previous publications.

Kluge et al. [70] found no correlation between the intensities of the 1066 meV PL signal and that of FeB pairs in EPR experiments after annealing the samples, and suggested that the PL centre is not due to B_s-Fe_i. No conclusion on the structure of the centre was reached, but a different configuration or the participation of other ions in the formation of the 1066 meV PL centre was suggested.

Davies et al. [71] explained the unusual pLVM structure of the 1066 meV centre, where the pLVM replicas at low temperature are stronger than the NP PL lines. The centre was reviewed by Istratov et al. [72], suggesting that EPR FeB centres might not have corresponding PL centres. It was suggested that sample contamination possibly led to incorrect conclusions in the past.

The 1066 meV PL centre was then observed in Au implanted Si [62], but only later was the involvement of Au confirmed by Henry et al. [50, 73] by observing the decay of the PL line with the decay of the ^{193}Au radioisotope. Although this method could not exclude any of the previously discussed constituents, evidence certainly increased against the necessity of Fe for the formation of this centre.

With the availability of isotopically purified ^{28}Si, it was then shown that the 1066 meV PL centre actually is Cu_4Au [45], whereas no Fe isotope shift could be observed. These experimental results are presented in detail in Sect. 4.2.3.

2.6 The 777 meV Cu$_4$Pt and the 884 meV Cu$_3$Pt Centres

Originally the 777 meV Cu_4Pt and the 884 meV Cu_3Pt centres were thought to be Fe related [2]. The 777 meV centre was described to have near isotropic and linear Zeeman splitting, which was attributed to cubic symmetry. However, later studies involving the ^{195}Au/^{195}Pt radioisotope decay pair found evidence that Pt played a major role in these centres [62]. At first, Henry et al. [62] ascribed the 777 meV line to the already documented 778 meV Ag system. After observing the rapid decay of this line in a sample implanted with ^{191}Pt and an increase in intensity in samples implanted with ^{195}Au decaying to ^{195}Pt, it was concluded that Pt and not Ag was responsible for this PL centre, overseeing the possibility of a potential overlap in energy in between two different centres. This overlap issue was resolved recently [45, 74]. In a later publication, Henry et al. [50, 73] confirmed the involvement of Pt in the 777 and 884 meV PL centres. The 777 meV centre was assigned to Pt-Fe initially and a decay of both PL lines was detected with the expected half-life time of the Pt radioisotope.

Alves et al. [75] have observed the 777 and 884 meV PL lines in samples implanted with natural Pt. Through preliminary uniaxial stress studies, it was concluded that the centres are axial in nature. A parallel was drawn between the 735 and 777 meV centres, as they were shown before to decay or increase in similar ratios in appropriate Au-Pt decay experiments [62]. Leitão et al. [76] have studied the 777 meV centre under uniaxial stress and determined it was an axial defect with C_{2v} symmetry. The centre was described as a donor-like IBE centre.

Only later has the involvement of 4, respectively 3 Cu atoms in the 777 and 884 meV centres in addition to Pt been shown [77] in high resolution PL in ^{28}Si. The presence of only a single Pt atom in these centres was confirmed in a further publication [74] and these results are explained in detail in Sect. 4.3. The centres were, therefore, labeled Cu$_4$Pt and Cu$_3$Pt respectively. The assumed connection between Cu$_4$Pt and Cu$_3$Au [75] could not be confirmed.

A 1026 meV PL centre that was mentioned on several occasions in literature [50, 73, 75], together with two previously discussed Pt defect centres, is generally too weak in our spectra to study its isotopic fingerprints. Alves et al. [75] claimed to observe a Si isotope effect and an intensity enhancement through Li, and concluded that the centre was likely large and involved a few atoms.

2.7 The 968 meV S$_A$ and 812 meV S$_B$ Centres

PL from S-doped Si was first reported in 1986 by Brown et al. [78]. Two PL systems were observed and labelled S$_B$ and S$_A$, with NP lines at $\sim$812 and $\sim$968 meV respectively [79]. A configurational metastability between S$_B$ and S$_A$ was observed and was the subject of several detailed studies [79–82]. A zero field ODMR study revealed a triplet splitting of both NP transitions [81, 83]. The detected nuclear hyperfine splitting due to a $I = 3/2$ spin suggested a single Cu at the core of both centres [81, 83]. Later, a model consisting of substitutional S and Cu$_i$ was proposed [82]. Both centres were found to have triclinic C_1 symmetry [82]. Bradfield et al. [84] reported a similar PL system in Si doped with Se, and suggested the involvement of O. It has recently been shown that the S$_B$ centre contains at least three Cu atoms in addition to S by analyzing the centre's isotopic fingerprint in ^{28}Si [44].

2.8 The 1044 meV "Q Line": The Li$_4$ V$_{Si}$ Centre

A discussion of this centre is included, even though it contains no TM, for two reasons. First, the properties of this centre have much in common with those of the previously discussed TM-containing IBE centres. Second, we believe that the structure, stability, and formation mechanisms of this centre are related to those of the TM centres, which are not yet fully understood. As shown later, Li can be a constituent in the TM-containing centres, which is perhaps not surprising given that both Li and TM are rapid interstitial diffusers in Si, and, being interstitials, are in a positive charge state. We speculate that the Li$_4$ V$_{Si}$ centre can be thought of as a member of the same family, where the V$_{Si}$ takes the place of a substitutional TM.

The 1044 meV PL centre was reported by Johnson et al. [85] and further invesitgated by Canham et al. [86, 87], who both concluded on the basis of temperature dependent, stress and isotope measurements that the "Q" lines at 1044, 1045, and 1048 meV thermalize with each other and are the electronic states of a centre with 4 Li atoms at a Si vacancy, with overall symmetry C_{3v}. In a subsequent study, DeLeo et al. [88] concluded that the Li atoms are likely situated in interstitial regions surrounding the vacancy, rather than placed in the vacancy itself. Lightowlers et al. [89] and Davies et al. [90] then confirmed previous experimental results, and found that decay time and Zeeman measurements indicate that the centre was a donor-like IBE. The properties of the Li$_4$ V$_{Si}$ centre were reviewed by Davies [8]. Later Tarnow [91] undertook *ab initio* calculations, and found that similar to earlier studies [88], the lowest energy ground state would have T_d tetrahedral symmetry with the Li atoms sitting interstitially outside of the Si vacancy, while not excluding an axial C_{3v} symmetry for the IBE state as observed in experiments.

It is interesting to note that although this centre contains Li, no high energy LVM replicas were observed, even though we will see that these do exist for Li-containing TM centres (see Sect. 4.3.4). The Li$_4$ V$_{Si}$ centre also has none of the very low energy pLVM replicas typical of the TM-containing IBE, which is perhaps easier to understand. The system does have some intermediate energy pLVM replicas, labelled *A, B,* and *C* [92], whose energies remain unexplained.

References

1. N. Minaev, A. Mudryi, V. Tkachev, Radiative recombination at thermal defects in silicon. Sov. Phys. Semicond. **13**, 233 (1979)
2. J. Weber, P. Wagner, Photoluminescence from Deep States associated with iron in silicon. J. Phys. Soc. Jpn. **49**(Suppl. A), 263–266 (1980)
3. J. Weber, H. Bauch, R. Sauer, Optical properties of copper in silicon: Excitons bound to isoelectronic copper pairs. Phys. Rev. B **25**, 7688–7699 (1982)
4. J.D. Struthers, solubility and diffusivity of gold, iron, and copper in silicon. J. Appl. Phys. **27**, 1560–1560 (1956)
5. K. Graff, H. Pieper, in Semiconductor Silicon, ed. by H.R. Ruff, R.S. Kriegler, Y. Takeishi (Electrochemical Society, Pennington, 1981)

6. S. Watkins, U. Ziemelis, M. Thewalt, Long lifetime photoluminescence from a deep centre in copper-doped silicon. Solid State Commun. **43**, 687–690 (1982)

7. R. Sauer, J. Weber, Vibronic coupling and implications in the copper related 1.014 eV photoluminescence spectrum in silicon. Solid State Commun. **49**, 833–836 (1984)

8. G. Davies, The optical properties of luminescence centres in silicon. Phys. Rep. **176**, 83–188 (1989)

9. S. Brotherton, J. Ayres, A. Gill, H. van Kesteren, F. Greidanus, Deep levels of copper in silicon. J. Appl. Phys. **62**, 1826–1832 (1987)

10. H.B. Erzgräber, K. Schmalz, Correlation between the Cu-related luminescent center and a deep level in silicon. J. Appl. Phys. **78**, 4066–4068 (1995)

11. M.H. Nazaré, A.J. Duarte, A.G. Steele, G. Davies, E.C. Lightowlers, Shallow excited states of the 1014 meV Cu related optical center in silicon. Mater. Sci. Forum **83–87**, 191–196 (1992)

12. A.A. Istratov, T. Heiser, H. Hieslmair, C. Flink, J. Krüger, E.R. Weber, A study of the copper-pair related centers in silicon. Mater. Sci. Forum **258–263**, 467–472 (1997)

13. A.A. Istratov, H. Hieslmair, T. Heiser, C. Flink, E.R. Weber, The dissociation energy and the charge state of a copper-pair center in silicon. Appl. Phys. Lett. **72**, 474–476 (1998)

14. S. Estreicher, Rich chemistry of copper in crystalline silicon. Phys. Rev. B **60**, 5375–5382 (1999)

15. S. Knack, J. Weber, H. Lemke, H. Riemann, Evolution of copper-hydrogen-related defects in silicon. Phys. B **308–310**, 404–407 (2001)

16. A.A. Istratov, E.R. Weber, Physics of copper in silicon. J. Electrochem. Soc. **149**, G21–G30 (2002)

17. S. Knack, Copper-related defects in silicon. Mater. Sci. Semicond. Process. **7**, 125–141 (2004)

18. M. Nakamura, S. Ishiwari, A. Tanaka, Number of Cu atom(s) in the 1.014 eV photoluminescence copper center and the center's model in silicon crystal. Appl. Phys. Lett. **73**, 2325–2327 (1998)

19. M. Nakamura, Dissociation of the 1.014 eV photoluminescence copper center in silicon crystal. Appl. Phys. Lett. **73**, 3896–3898 (1998)

20. M. Nakamura, H. Iwasaki, Copper complexes in silicon. J. Appl. Phys. **86**, 5372–5375 (1999)

21. M. Nakamura, S. Murakami, Depth progression of dissociation reaction of the 1.014 eV photoluminescence copper center in copper-diffused silicon crystal measured by deep-level transient spectroscopy. Appl. Phys. Lett. **98**, 141909 (2011)

22. S. Estreicher, D. West, J. Pruneda, S. Knack, J. Weber, Formation and properties of three copper pairs in silicon. Mater. Res. Soc. Symp. Proc. **719**, 421–426 (2002)

23. S. Estreicher, D. West, J. Goss, S. Knack, J. Weber, First-principles calculations of pseudolocal vibrational modes: the case of Cu and Cu pairs in Si. Phys. Rev. Lett. **90**, 035504 (2003)

24. S. Estreicher, First-principles theory of copper in silicon. Mater. Sci. Semicond. Process. **7**, 101–111 (2004)

25. M. Nakamura, H. Ohno, S. Murakami, Formation of the 1.014 eV photoluminescence Cu center in Cu-implanted silicon crystals and the center's model. Jpn. J. Appl. Phys. **43**, L1466–1468 (2004)

26. M. Nakamura, S. Murakami, H. Hozoji, N. Kawai, S. Saito, H. Arie, Concentration study of deep-level Cu center in Cu-diffused Si crystals by deep-level transient spectroscopy and photoluminescence measurements. Jpn. J. Appl. Phys. **45**, L80 (2006)

27. M. Thewalt, M. Steger, A. Yang, M. Cardona, H. Riemann, N. Abrosimov, M. Churbanov, A. Gusev, A. Bulanov, I. Kovalev, A. Kaliteevskii, O. Godisov, P. Becker, H.-J. Pohl, Can highly enriched ^{28}Si reveal new things about old defects? Phys. B **401–402**, 587–592 (2007)

28. D. Karaiskaj, M. Thewalt, T. Ruf, M. Cardona, H.-J. Pohl, G. Deviatych, P. Sennikov, H. Riemann, Photoluminescence of isotopically purified silicon: how sharp are bound exciton transitions? Phys. Rev. Lett. **86**, 6010–6013 (2001)

29. D. Karaiskaj, J. Stotz, T. Meyer, M. Thewalt, M. Cardona, Impurity absorption spectroscopy in ^{28}Si: the importance of inhomogeneous isotope broadening. Phys. Rev. Lett. **90**, 186402 (2003)

30. M. Cardona, M. Thewalt, Isotope effects on the optical spectra of semiconductors. Rev. Mod. Phys. **77**, 1173 (2005)
31. A. Yang, M. Steger, D. Karaiskaj, M. Thewalt, M. Cardona, K. Itoh, H. Riemann, N. Abrosimov, M. F. Churbanov, A. Gusev, A. Bulanov, A. Kaliteevskii, O. Godisov, P. Becker, H.-J. Pohl, J. Ager III, E. Haller, Optical detection and ionization of donors in specific electronic and nuclear spin states. Phys. Rev. Lett. **97**, 227401 (2006)
32. M. Thewalt, A. Yang, M. Steger, D. Karaiskaj, M. Cardona, H. Riemann, N. Abrosimov, A. Gusev, A. Bulanov, I. Kovalev, A. Kaliteevskii, O. Godisov, P. Becker, H. Pohl, E. Haller, J. Ager III, K. Itoh, Direct observation of the donor nuclear spin in a near-gap bound exciton transition: ^{31}P in highly enriched ^{28}Si. J. Appl. Phys. **101**, 081724 (2007)
33. M. Nakamura, S. Murakami, N.J. Kawai, S. Saito, H. Arie, Diffusion-temperature-dependent formation of Cu centers in Cu-saturated silicon crystals studied by photoluminescence and deep-level transient spectroscopy. Jpn. J. Appl. Phys. **47**, 4398–4402 (2008)
34. M. Nakamura, S. Murakami, N.J. Kawai, S. Saito, K. Matsukawa, H. Arie, Compositional transformation between Cu centers by annealing in Cu-diffused silicon crystals studied with deep-level transient spectroscopy and photoluminescence. Jpn. J. Appl. Phys. **48**, 082302 (2009)
35. M. Nakamura, S. Murakami, Deep-level transient spectroscopy and photoluminescence studies of formation and depth profiles of copper centers in silicon crystals diffused with dilute copper. Jpn. J. Appl. Phys. **49**, 071302 (2010)
36. K. Shirai, H. Yamaguchi, A. Yanase, H. Katayama-Yoshida, A new structure of Cu complex in Si and its photoluminescence. J. Phys. Cond. Mater. **21**, 064249 (2009)
37. K. Shirai, H. Yamaguchi, J. Ishisada, K. Matsukawa, A. Yanase, S. Emura, Cu complex in silicon and its photoluminescence, in *AIP Conference Proceedings*, vol. 1199, ed. by M. Caldas, N. Studart, 2010, pp. 91–92
38. S.K. Estreicher, A. Carvalho, The Cu_{PL} defect and the $Cu_{s1}Cu_{i3}$ complex. Phys. B: Cond. Mat. **407**(15), 2967–2969 (2012)
39. A. Carvalho, D.J. Backlund, S.K. Estreicher, Four-copper complexes in Si and the Cu_{PL} defect: a first-principles study. Phys. Rev. B **84**, 155322 (2011)
40. K. McGuigan, M. Henry, E. Lightowlers, A. Steele, M. Thewalt, A new photoluminescence band in silicon lightly doped with copper. Solid State Commun. **68**, 7–11 (1988)
41. K. McGuigan, M. Henry, M. Carmo, G. Davies, E. Lightowlers, A uniaxial stress study of a copper-related photoluminescence band in silicon. Mater. Sci. Eng. B **4**, 269–272 (1989)
42. S. Estreicher, D. West, M. Sanati, $^{*}Cu_{0}$: a metastable configuration of the Cu_{s}-Cu_{i} pair in Si. Phys. Rev. B **72**, 121201 (2005)
43. N.Q. Vinh, J. Phillips, G. Davies, T. Gregorkiewicz, Time-resolved free-electron laser spectroscopy of a copper isoelectronic center in silicon. Phys. Rev. B **71**, 085206 (2005)
44. A. Yang, M. Steger, M. Thewalt, M. Cardona, H. Riemann, N. Abrosimov, M. Churbanov, A. Gusev, A. Bulanov, I. Kovalev, A. Kaliteevskii, O. Godisov, P. Becker, H.-J. Pohl, J. Ager III, E. Haller, High resolution photoluminescence of sulphur- and copper-related isoelectronic bound excitons in highly enriched ^{28}Si. Phys. B **401–402**, 593–596 (2007)
45. M. Steger, A. Yang, N. Stavrias, M. Thewalt, H. Riemann, N. Abros imov, M. Churbanov, A. Gusev, A. Bulanov, I. Kovalev, A. Kaliteevskii, O. Godisov, P. Becker, H.-J. Pohl, Reduction of the linewidths of deep luminescence centers in ^{28}Si reveals fingerprints of the isotope constituents. Phys. Rev. Lett. **100**, 177402 (2008)
46. N.Q. Vinh, T. Gregorkiewicz, Two-color mid-infrared spectroscopy of isoelectronic centers in silicon, in *MRS Proceedings*, vol. 770 (Cambridge University Press, Cambridge, 2003), p. I4.2
47. J. Olajos, M. Kleverman, H. Grimmeiss, High-resolution spectroscopy of silver-doped silicon. Phys. Rev. B **38**, 10633–10640 (1988)
48. N. Baber, H.G. Grimmeiss, M. Kleverman, P. Omling, M. Zafar Iqbal, Characterization of silver-related deep levels in silicon. J. Appl. Phys. **62**, 2853–2857 (1987)
49. M. Nazaré, M. Carmo, A. Duarte, Luminescence from transition metal centres in silicon doped with silver and nickel. Mater. Sci. Eng. B **4**, 273–276 (1989)
50. M.O. Henry, E. Alves, J. Bollmann, A. Burchard, M. Deicher, M. Fanciulli, D. Forkel-Wirth, M.H. Knopf, S. Lindner, R. Magerle, R. McGlynn, K.G. McGuigan, J.C. Soares, A. Stotzler, G.

Weyer, Radioactive isotope identifications of Au and Pt photoluminescence centres in silicon. Phys. Status Solidi B **210**, 853 (1998)

51. N.T. Son, M. Singh, J. Dalfors, B. Monemar, E. Janzén, Electronic structure of a photoluminescent center in silver-doped silicon. Phys. Rev. B **49**, 17428–17431 (1994)

52. M. Singh, W. Chen, N. Son, B. Monemar, E. Janzén, Shallow excited states of deep luminescent centers in silicon. Solid State Commun. **93**, 415–418 (1995)

53. N.T. Son, V.E. Kustov, T. Gregorkiewicz, C.A.J. Ammerlaan, Electron-paramagnetic-resonance identification of silver centers in silicon Phys. Rev. B **46**, 4544–4550 (1992)

54. N.T. Son, T. Gregorkiewicz, C.A.J. Ammerlaan, Magnetic resonance spectroscopy in silver-doped silicon. J. Appl. Phys. **73**, 1797–1801 (1993)

55. M.Z. Iqbal, G. Davies, E.C. Lightowlers, Photoluminescence from silver-related defects in silicon, in *Defects in Semiconductors*, ICDS-17, vol. 143, ed. by H. Heinrich, W. Jantsch, Materials Science Forum (1994), pp. 773–777

56. N. Vinh, T. Gregorkiewicz, K. Thonke, 780-meV photoluminescence band in silver-doped silicon: isotope effect and time-resolved spectroscopy. Phys. Rev. B **65**, 033202 (2001)

57. G. Davies, T. Gregorkiewicz, M.Z. Iqbal, M. Kleverman, E. Lightowlers, N. Vinh, M. Zhu, Optical properties of a silver-related defect in silicon. Phys. Rev. B **67**, 235111 (2003)

58. J.H. Svensson, B. Monemar, E. Janzén, Pseudodonor electronic excited states of neutral complex defects in silicon. Phys. Rev. Lett. **65**, 1796–1799 (1990)

59. K. Otha, Infrared absorption of gold in silicon single crystal. Sci. Light **22**, 12 (1973)

60. T. Schlesinger, R. Hauenstein, R. Feenstra, T. McGill, Isotope shifts for the P, Q, R lines in indium-doped silicon. Solid State Commun. **46**, 321–324 (1983)

61. M. do Carmo, M. Calão, G. Davies, E. Lightowlers, Photoluminescence from transition metals in silicon, in *Defects in Semiconductors* (Trans Tech, Aedermannsdorf, 1989), p. 1497

62. M. Henry, S. Daly, C. Frehill, E. McGlynn, C. McDonagh, A photoluminescence study of gold- and platinum-related defects in silicon using radioactive transformations, in *Physics of Semiconductors: 23rd International Conference on the Physics of Semiconductors-ICPS*, ed. by M. Scheffler, R. Zimmermann (1996), pp. 2713–2716

63. M. Henry, E. McGlynn, J. Fryar, S. Lindner, J. Bollmann, The evolution of point defects in semiconductors studied using the decay of implanted radioactive isotopes. Nucl. Instrum. Methods Phys. Res. B **178**, 256–259 (2001)

64. R. Sauer, J. Weber, Photoluminescence characterisation of deep defects in silicon. Phys. B+C **116**, 195 (1983)

65. K. Wünstel, P. Wagner, Interstitial iron and iron-acceptor pairs in silicon. Appl. Phys. A **27**, 207–212 (1982)

66. K. Graff, H. Pieper, The properties of iron in silicon. J. Electrochem. Soc. **128**, 669–674 (1981)

67. T.E. Schlesinger, T.C. McGill, Isotope-shift experiments on luminescence attributed to (Fe, B) pairs in silicon. Phys. Rev. B **28**, 3643–3644 (1983)

68. H.D. Mohring, J. Weber, R. Sauer, Photoluminescence of excitons bound to an isoelectronic trap in silicon associated with boron and iron. Phys. Rev. B **30**, 894–904 (1984)

69. H. Conzelmann, Photoluminescence of transition metal complexes in silicon. Appl. Phys. A **42**, 1–18 (1987)

70. J. Kluge, W. Gehlhoff, J. Donecker, Combined EPR and luminescence investigations of centers in silicon associated with boron and iron. Acta Phys. Pol. A **73**, 207 (1988)

71. G. Davies, M.C. do Carmo, Vibronic coupling in shallow excited states of optical centres in silicon. Inst. Phys. Conf. Ser. **95**, 125–130 (1989)

72. A.A. Istratov, H. Hieslmair, E.R. Weber, Iron and its complexes in silicon. Appl. Phys. A **69**, 13–44 (1999)

73. M. Henry, M. Deicher, R. Magerle, E. McGlynn, A. Stotzler, Photoluminescence analysis of semiconductors using radioactive isotopes. Hyperfine Interact. **129**, 443–460 (2000)

74. M. Steger, A. Yang, T. Sekiguchi, K. Saeedi, M.L.W. Thewalt, M. O. Henry, K. Johnston, E. Alves, U. Wahl, H. Ricmann, N.V. Abrosimov, M.F. Churbanov, A.V. Gusev, A.K. Kaliteevskii, O.N. Godisov, P. Becker, H.-J. Pohl, Isotopic fingerprints of Pt-containing luminescence centers in highly enriched ^{28}Si. Phys. Rev. B **81**, 235217 (2010)

75. E. Alves, J. Bollmann, M. Deicher, M.C. Carmo, M.O. Henry, M.H.A. Knopf, J.P. Leitao, R. Magerle, C.J. McDonagh, The photoluminescence of Pt-implanted silicon, in *Defects in Semiconductors—ICDS-19*, vol. 258–2, ed. by G. Davies, M.H. Nazare, Materials Science Forum (1997), pp. 473–478

76. J.P. Leitão, M.C. Carmo, M.O. Henry, E. McGlynn, J. Bolmann, S. Lindner, The 777 meV photoluminescence band in Si:Pt. Phys. B **273–274**, 420–423 (1999)

77. M. Steger, A. Yang, T. Sekiguchi, K. Saeedi, M. Thewalt, M. Henry, K. Johnston, H. Riemann, N. Abrosimov, M. Churbanov, A. Gusev, A. Bulanov, I. Kaliteevski, O. Godisov, P. Becker, H.-J. Pohl, Isotopic fingerprints of gold-containing luminescence centers in ^{28}Si. Phys. B **404**, 5050–5053 (2009)

78. T.G. Brown, D.G. Hall, Optical emission at 1.32 μm from sulfur-doped crystalline silicon. Appl. Phys. Lett. **49**, 245–247 (1986)

79. D.J.S. Beckett, M.K. Nissen, M.L.W. Thewalt, Optical properties of the sulfur-related isoelectronic bound excitons in Si. Phys. Rev. B **40**, 9618–9625 (1989)

80. M. Singh, E.C. Lightowlers, G. Davies, C. Jeynes, K.J. Reeson, Isoelectronic bound exciton photoluminescence from a metastable defect in sulphur-doped silicon. Mater. Sci. Eng. B **4**, 303–307 (1989)

81. A.M. Frens, M.T. Bennebroek, J. Schmidt, W.M. Chen, B. Monemar, Zero-field optical detection of magnetic resonance on a metastable sulfur-pair-related defect in silicon: Evidence for a Cu constituent. Phys. Rev. B **46**, 12316–12322 (1992)

82. P.W. Mason, H.J. Sun, B. Ittermann, S.S. Ostapenko, G.D. Watkins, L. Jeyanathan, M. Singh, G. Davies, E.C. Lightowlers, Sulfur-related metastable luminescence center in silicon. Phys. Rev. B **58**, 7007–7019 (1998)

83. A.M. Frens, M.E. Braat, J. Schmidt, W.M. Chen, B. Monemar, Transfer mechanism between pseudodonor excited singlet and triplet states of the S-Cu complex defect in silicon. Phys. Rev. B **52**, 8848–8853 (1995)

84. P.L. Bradfield, T.G. Brown, D.G. Hall, Radiative decay of excitons bound to chalcogen-related isoelectronic impurity complexes in silicon. Phys. Rev. B **38**, 3533–3536 (1988)

85. E.S. Johnson, W.D. Compton, J.R. Noonan, B.G. Streetman, Recombination luminescence from electron-irradiated Li-diffused Si. J. Appl. Phys. **44**, 5411–5418 (1973)

86. L. Canham, G. Davies, E. Lightowlers, The 1.045 eV vibronic band in silicon doped with lithium. J. Phys. C Solid State Phys. **13**, L757 (1980)

87. L. Canham, G. Davies, E. Lightowlers, The 1.045 eV vibronic band in irradiated silicon doped with lithium. Inst. Phys. Conf. Ser. **59**, 211–216 (1981)

88. G.G. DeLeo, W.B. Fowler, G.D. Watkins, Electronic structure of hydrogen- and alkali-metal-vacancy complexes in silicon. Phys. Rev. B **29**, 1819–1823 (1984)

89. E.C. Lightowlers, L.T. Canham, G. Davies, M.L.W. Thewalt, S.P. Watkins, Lithium and lithium-carbon isoelectronic complexes in silicon: luminescence-decay-time, absorption, isotope-splitting, and Zeeman measurements. Phys. Rev. B **29**, 4517–4523 (1984)

90. G. Davies, L. Canham, E. Lightowlers, Magnetic and uniaxial stress perturbations of optical transitions at a four Li atom complex in Si. J. Phys. C **17**, L173–L178 (1984)

91. E. Tarnow, Theory of Li- and Br-saturated vacancies in radiation damaged silicon. J. Phys. Condens. Mater. **4**, 1459–1464 (1992)

92. L. Canham, G. Davies, E. Lightowlers, G. Blackmore, Complex isotope splitting of the no-phonon lines associated with exciton decay at a four-lithium-atom isoelectronic centre in silicon. Phys. B **117–118**, 119–121 (1983)

Chapter 3
Experimental Method

This chapter presents details about the samples and the experimental setup. The first section discusses the materials used to prepare the samples and explains the process of sample preparation. This is followed by a section giving an overview of Fourier transform spectroscopy, and ends with an explanation of the experimental setup and technical means used to carry out the photoluminescence experiments.

3.1 Silicon Samples

3.1.1 Sample Preparation

All of the results presented in this work use the same highly enriched (99.991 %) ^{28}Si material, which has produced the narrowest linewidths in our previous studies [1–6] and originates from the Avogadro project [7–9]. This p-type ^{28}Si crystal has a P concentration of $\sim 2 \times 10^{12}\,\mathrm{cm}^{-3}$ and a B concentration of $\sim 5 \times 10^{13}\,\mathrm{cm}^{-3}$, as measured by photoluminescence, and a C concentration of $<5 \times 10^{14}\,\mathrm{cm}^{-3}$ (detection limit) as measured by local vibrational mode absorption [10]. To study the many IBE centres reported in this work, small samples of the ^{28}Si crystal were treated with different TM and other impurities to obtain the desired optical centres. After cutting them from the parent crystal, the samples were etched in $\mathrm{HF/HNO_3}$ to remove surface damage, and then diffused with the desired impurities which were introduced either from surface layers, from the gaseous phase in sealed quartz ampoules, or by ion implantation followed by annealing.

Normally, diffusion and annealing steps were carried out in sealed ampoules, after evacuation and backfilling with Ar gas at temperatures and diffusion times described in previous studies of the same IBE centres. The most commonly used diffusion conditions are listed in Table 3.1. Diffusants were either evaporated onto the sample before sealing it in the ampoule or a small quantity was sealed in the ampoule together with the sample. After heat treatment, the sealed ampoules were

M. Steger, *Transition-Metal Defects in Silicon*, Springer Theses,
DOI: 10.1007/978-3-642-35079-5_3, © Springer-Verlag Berlin Heidelberg 2013

Table 3.1 Typical annealing conditions for Pt and Au implants and diffusion conditions for Au, Ag, Cu, S, and Li

	T (°C)	t		Method
Au	1000	10 min	Tube	(Annealing)
Pt	950	10 min	Tube	(Annealing)
Au	950	2 h	Ampoule	(Diffusion)
Pt	1000	24 h	Ampoule	(Diffusion)
Ag	950	2 h	Ampoule	(Diffusion)
Cu	800	2 min	Tube	(Diffusion)
S	1100	20 h	Ampoule	(Diffusion)
Li	800	2 min	Tube	(Diffusion)

quenched directly into methanol, resulting in a moderately fast quench rate. After this, the samples were removed from the ampoules, cleaned, and etched in HF/HNO_3 to ensure a damage-free surface. The importance of quenching for the formation of IBE centres was discovered earlier [11, 12]. Quenching prevents the highly mobile Cu_i from out-diffusing [13], which would happen during a slow cool-down process. In some cases, the samples were requenched later directly into methanol after a quick heating to $\sim$700 °C to achieve a much higher cooling rate. While this sealed ampoule technique was used originally for Cu diffusions, it was later discovered that isotope-specific Cu and Li diffusions could be accomplished more quickly by heating the sample for several minutes under flowing Ar in an open quartz tube whose inner surface was coated intentionally with the desired species, and then quenching that sample into methanol.

Sample surfaces and quartzware were cleaned with KCN solution before any diffusion or annealing treatments to remove metallic contaminants. While this method generally works well, trace amounts of Cu were always detected in heat-treated and quenched samples. Cu is an essential part of many of the discussed PL centres. This residual Cu contamination could be seen mainly in the formation of complexes other than Cu_4, which was always of low intensity without intentional Cu diffusion. While this unintentional Cu accurately reflected the ^{nat}Cu abundances before enriched Cu diffusions were performed, later it also showed slightly different ratios due to contamination of the furnace and quartzware. In some cases, such as during the formation of the Ag_4 centre, it was not desirable to trap the Cu in the sample because that would lead to the formation of Cu-Ag complexes at the cost of Ag_4. The formation of the Ag_4 centres was aided by slow cooling or low temperature annealing ($\sim$100 °C), thereby reducing the amount of available Cu_i.

The formation of the Li related centres was observed to be a slow process. The more Li a given centre contained, the less likely it formed during the initial sample quenches, after which Cu-rich centres dominated the PL spectrum. Storing the sample for several days at room temperature significantly increased the intensity of Li centres. This process was accelerated by low temperature annealing (for example 60 °C over several days), and, even after longer storage times of several months an increase in the intensity of many Li-rich PL lines was observed. During this process, the highly mobile Cu_i likely diffused out of the bulk, being replaced by Li that remains at a high concentration within the sample due to its lower diffusion coefficient at room temperature. Therefore, even at room temperature, these centres may not be truly stable.

For this work, the impurities diffused into the samples were high purity, natural Cu, Li, S, Ag, Pt, and Au. The natural isotope abundances of these elements are provided in Table 3.2. To better understand the isotopic fingerprint structure of the observed complexes, the following high purity, enriched single isotopes were used in addition: 99.8 % ^{63}Cu; 99.6 % ^{65}Cu; 95.5 % ^{6}Li; 99.99 % ^{7}Li; 99 % ^{107}Ag; and 99 % ^{109}Ag.

The problem of Au having only one stable isotope was solved by preparing samples containing the radioisotope ^{195}Au. Discs of ^{28}Si with 1 mm thickness were implanted with ^{195}Hg at 60 keV using the ISOLDE facility at CERN [6, 14–16]. The ^{195}Hg decayed rapidly to ^{195}Au, and this conversion was completed before the PL experiments were started. One set of samples was implanted with only ^{195}Hg/^{195}Au at a dose of 2×10^{13} cm^{-2}; another set was implanted with 1×10^{13} cm^{-2} ^{195}Hg/^{195}Au and the same target dose of ^{197}Au (at 100 keV); whereas other samples were prepared to contain only ^{197}Au either by implantation (1×10^{15} cm^{-2} at 100 keV) or by diffusion from a ^{197}Au surface layer. ^{195}Au decays to stable ^{195}Pt with a half life of 186 days, therefore, over time, these samples contained both a decreasing concentration of Au and an increasing concentration of Pt, which allowed for the study of Pt-containing centres in the samples originally intended for Au studies. The decay of ^{195}Au involves only a small recoil energy of 0.3 eV and it was expected that no site change of the decaying atom would take place [17]. ^{195}Pt is believed to occupy substitutional lattice sites [18] and, subsequently, this is expected to hold for ^{195}Au as well [17]. For comparison purposes, other Pt-containing samples were prepared by implanting ^{194}Pt, ^{195}Pt, and ^{198}Pt at 60 keV with a dose of approximately 10^{14} cm^{-2} and one by in-diffusion of ^{nat}Pt. The implanted samples and quartzware were cleaned in a KCN solution to remove metallic surface contaminants and annealed at 900–1000 °C for 10–30 min under flowing Ar to both remove the implantation damage, and to allow the Au or Pt to diffuse into the bulk of the sample, and then quenched to room temperature in methanol.

Table 3.2 Natural isotopic abundances of the impurities used in this work in percent

^{nat}Cu	^{nat}Li	^{nat}S	^{nat}Ag	^{nat}Pt	^{nat}Au
^{63}Cu 69.2	^{6}Li 7.6	^{32}S 95.0	^{107}Ag 51.8	^{190}Pt 0.01	^{197}Au 100.0
^{65}Cu 30.8	^{7}Li 92.4	^{33}S 0.8	^{109}Ag 48.2	^{192}Pt 0.8	
		^{34}S 4.3		^{194}Pt 33.0	
		^{36}S 0.02		^{195}Pt 33.8	
				^{196}Pt 25.2	
				^{198}Pt 7.2	

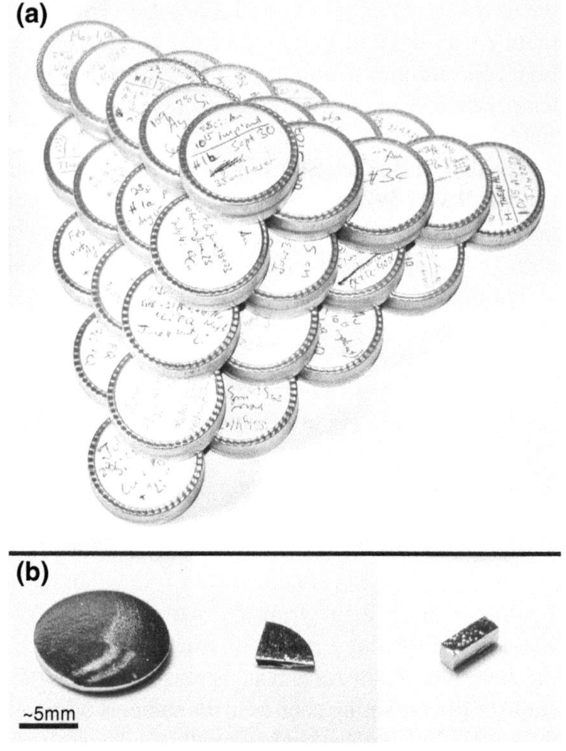

Fig. 3.1 **a** Some of the samples used for this research project. **b** A ^{28}Si disc after ion-implantation, and a quarter piece of a similar disc, and a ^{28}Si stick, both after going through diffusion processes. All samples were etched in HF/HNO$_3$ to remove surface damage. The scale bar indicates the approximate size of all samples

In total, approximately 50 samples were produced in this way, including some ^{nat}Si control samples. Most of the ^{28}Si samples were quarter pieces of 1 mm thick discs; only a few have rectangular shape, as shown in Fig. 3.1. Typically most samples have undergone several treatments to obtain different combinations of impurity isotopes.

3.1.2 Cu Diffusion

The diffusion process used to prepare the samples in the ampoule or tube can be described as a constant source diffusion, because a sufficiently high amount of Cu is present in the ampoule. In most cases, the samples were coated with metal before sealing them into the ampoule. The depth and time dependent diffusion profile can be calculated according to [19]:

$$N(x, t) = N_0 \operatorname{erfc}\left(\frac{x}{2\sqrt{Dt}}\right) \qquad (3.1)$$

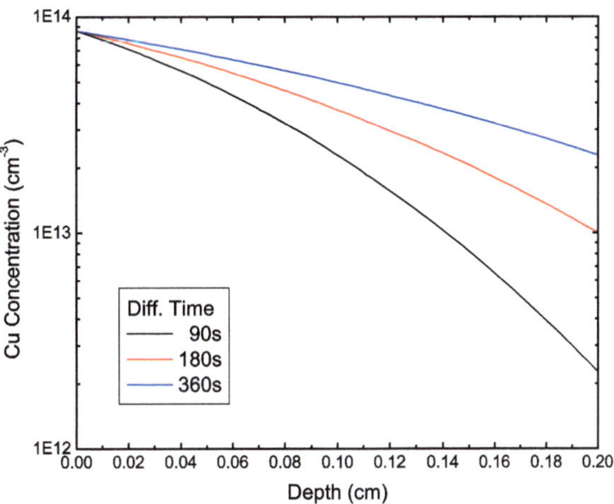

Fig. 3.2 The constant source diffusion profile of a Cu diffusion for different diffusion times t with a diffusion coefficient $D = 4.5 \times 10^{-5}\,\mathrm{cm^2/s}$ (at $800\,^\circ\mathrm{C}$)

where N_0 is the impurity concentration at the surface and D is the diffusion coefficient. To use this equation, we assume a constant source of Cu in the quartz tube. At a diffusion temperature of $800\,^\circ\mathrm{C}$, we obtain a diffusion coefficient of $D = 4.5 \times 10^{-5}\,\mathrm{cm^2/s}$ according to Eq. (1.14). A sample thickness of $x = 2\,\mathrm{mm}$ is assumed. The diffusion time t is typically several minutes long; for this calculation we choose $t = 180\,\mathrm{s}$. It follows, that to obtain a diffused Cu concentration of $N(x,t) = 10^{13}\,\mathrm{cm^{-3}}$, a surface concentration of $N_0 = 8.6 \times 10^{13}\,\mathrm{cm^{-3}}$ is needed. This is a value that is easily satisfied by a thin Cu vapor deposition. With a density of $\sim 10^{22}\,\mathrm{cm^{-3}}$, a thin Cu layer can be considered as a constant source for this purpose. Therefore the solubility limit of Cu in Si can be reached under these diffusion conditions. Figure 3.2 shows the diffusion profile for three different diffusion times.

3.1.3 Au and Pt Ion Implantation

Different isotopic species of Au and Pt were implanted under similar conditions. Figure 3.3 shows a simulation of a typical Au implantation into Si. After the implantation step, the samples were cleaned and annealed. The annealing process heals implantation damage and also acts as a limited source diffusion. The profile of a limited source diffusion can be calculated according to [19]:

$$N(x,t) = \frac{Q}{\sqrt{\pi D t}} \times e^{-(x/2\sqrt{Dt})^2} \tag{3.2}$$

where Q is the implantation dose. The ^{197}Au implantation has a dose of $Q = 10^{15}\,\mathrm{cm^{-2}}$ and the Au diffusion coefficient is $D = 4 \times 10^{-7}\,\mathrm{cm^2/s}$ at $1000\,^\circ\mathrm{C}$ [19]. The Au concentration after a $t = 10\,\mathrm{min}$ diffusion process under

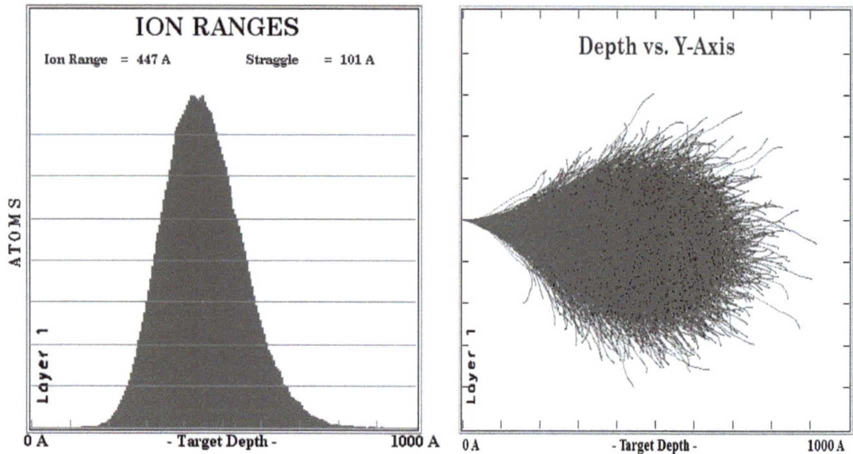

Fig. 3.3 Simulation of ion ranges and spread for an implantation of [nat]Au at 100 keV and 7° tilt to the [001] normal of [nat]Si. Created with the SRIM software package [20]

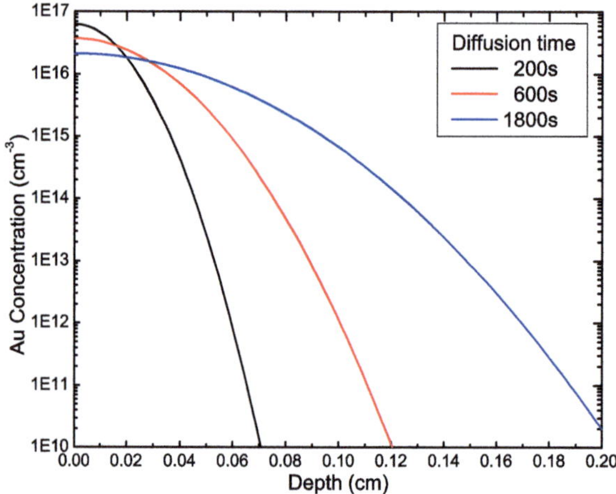

Fig. 3.4 The limited source diffusion profile of an Au implantation with a dose $Q = 10^{15}\,\text{cm}^{-2}$ after 10 min at 1000 °C ($D = 4 \times 10^{-7}\,\text{cm}^2/\text{s}$) for three different diffusion times

these conditions is approximately $10^{12}\,\text{cm}^{-3}$ at a depth of 1 mm. Figure 3.4 shows the diffusion profile for three different diffusion times.

3.2 Fourier Transform Spectroscopy

Typically as in this case, a Fourier transform spectrometer is a Michelson interferometer, originally designed in 1891 [21, 22]. The Michelson interferometer splits the incoming beam into two, and then recombines those beams after introducing a

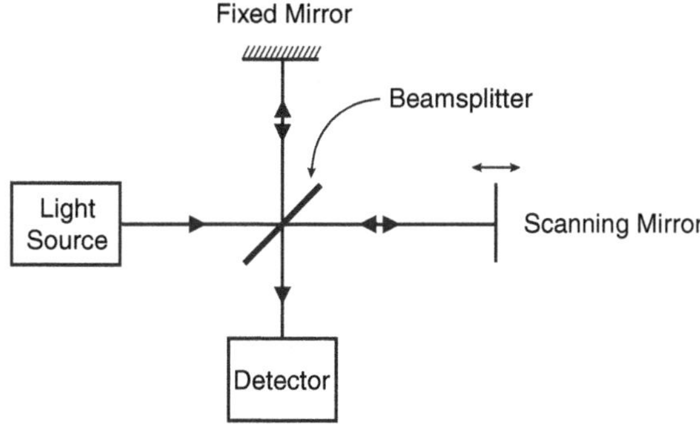

Fig. 3.5 Schematic of a Michelson interferometer

variable path difference to one of the split beams by means of a scanning mirror (see Fig. 3.5). Then the output signal can be recorded as a variable of the path difference δ, which is also referred to as retardation.

Given this geometry, a monochromatic light source would interfere with itself in the detector plain due to the path difference of the two beams, thereby resulting in the following retardation dependent intensity $I'(\delta)$ [23]:

$$I'(\delta) = 0.5 I(v)(1 + \cos 2\pi v\delta) \tag{3.3}$$

where $v = 1/\lambda$ is the wavenumber and $I(v)$ is the source intensity. The constant component does not contain any relevant information. The modulated intensity $I'(\delta)$ is referred to as the interferogram. In a real world spectrometer, other factors such as the beam splitter, the detector, and the amplifier modify the signal. These factors can be modelled with a single wavenumber dependent factor $H(v)$. Setting $B(v) = 0.5 H(v) I(v)$ gives for the interferogram:

$$I(\delta) = B(v) \cos 2\pi v\delta \tag{3.4}$$

where $I(\delta)$ is the cosine Fourier transform (FT) of $B(v)$. To obtain the spectrum $B(v)$ the FT of the interferogram $I(\delta)$ has to be calculated.

When we take Eq. (3.4) from a monochromatic to a continuous source, the interferogram is given by an integral over the wavenumber:

$$I(\delta) = \int_{-\infty}^{+\infty} B(v) \cos 2\pi v\delta \, dv \tag{3.5}$$

with its cosine Fourier transform being:

$$B(v) = \int_{-\infty}^{+\infty} I(\delta) \cos 2\pi v\delta \, d\delta \tag{3.6}$$

Here we see that it is possible to obtain a scan of the entire spectrum over all wavenumbers, but on the other hand Eq. (3.6) also shows a limitation of the Fourier transform spectrometer. A real interferometer will always have a limited retardation δ and as we shall see, this results in a finite resolution.

The maximum retardation of the interferogram is restricted to a finite length Δ. A simple truncation function $D(\delta)$, also called the boxcar function, would be:

$$D(\delta) = \begin{cases} 1 & \text{if } -\Delta \le \delta \le +\Delta \\ 0 & \text{if } \delta > |\Delta| \end{cases} \tag{3.7}$$

The spectrum is then modified so that:

$$B(\nu) = \int_{-\infty}^{+\infty} I(\delta)D(\delta) \cos 2\pi \nu \delta \, d\delta \tag{3.8}$$

Hence the spectrum is a convolution of the FT of $I(\delta)$ and $D(\delta)$. Whereas the FT of $I(\delta)$ results in the true spectrum, the FT of $D(\delta)$ is:

$$f(\nu) = 2\Delta \, \text{sinc} \, 2\pi \nu \Delta \tag{3.9}$$

that is a sinc-function, shown in Fig. 3.6.

When a true line of infinitesimal width is convoluted with a sinc function, its width turns finite. For a boxcar truncation function, the full width at half maximum (FWHM) imposed by the finite path length is [24]:

$$\text{FWHM} = \frac{1.207}{2\Delta} \tag{3.10}$$

The maximum retardation in the Bomem DA8 spectrometer used for this study is $\Delta = 250$ cm, corresponding to a maximum resolution of $0.0024\,\text{cm}^{-1}$ $(0.3\,\mu\text{eV})$ using the boxcar truncation.

Due to its shape, the sinc function introduces "ringing" in the spectrum surrounding sharp features that are narrower than the resolution used. Using a different

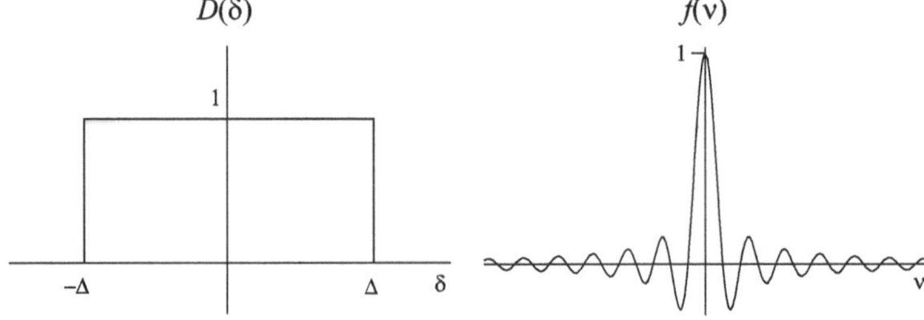

Fig. 3.6 Plot of the $D(\delta)$ boxcar function and its Fourier transform $f(\nu) \sim \text{sinc} \, \nu$

apodization function for the truncation of the interferogram, like the Blackman-Harris function, these artifacts can be reduced at the cost of increased FWHM [23].

3.3 Experimental Setup

3.3.1 Fourier Transform Spectrometer

In this study, a Bomem DA8 Fourier transform spectrometer was used with a CaF_2 beam splitter to obtain the PL spectra. The instrumental resolution used was between 2.5 and 0.6 µeV for high resolution spectra, and 60 µeV for low resolution spectra (as used for spectra of LVM and pLVM replicas).

The PL spectra were collected using bulk excitation at 1047 nm from a Nd:YLF laser and a liquid nitrogen cooled Ge photodetector (North Coast Scientific E0-817S), or an InSb detector for spectral regions below the Ge cut-on ($\sim$750 meV). The laser beam was directed onto the sample in the cryostat, so that no laser light was directly reflected into the collection path of the spectrometer. In some cases doing this was not possible and two laser rejection filters (Semrock 1064LP), one in the collection path and the other one between aperture and detector, were used in addition to a 780LP filter. Figure 3.7 shows a schematic diagram of the spectrometer and its peripherals.

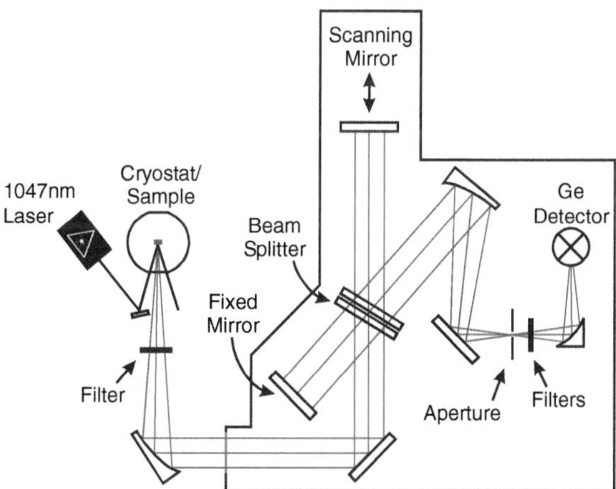

Fig. 3.7 Experimental setup with Fourier-transform interferometer, cryostat, laser, and detector. The laser beam was directed onto the sample in the cryostat, so that no laser light was reflected into the collection path of the spectrometer. The emitted photoluminescence signal was then collected with the spectrometer, where the light passes through the actual interferometer with its two arms, before being passed through a circular aperture and focused onto the detector

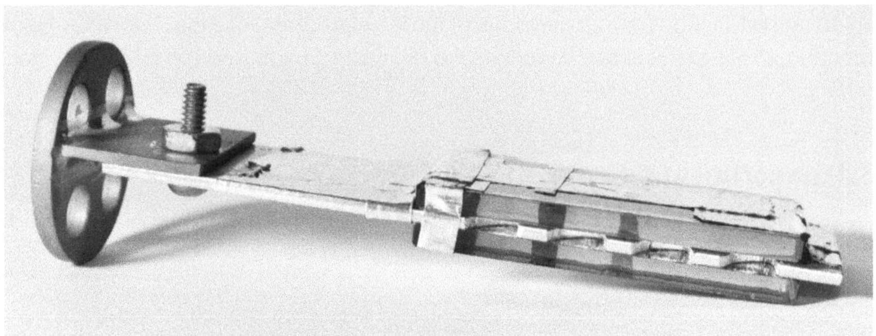

Fig. 3.8 A sample holder used to submerge the samples is superfluid He. The samples fit into the quarter-disc sized cavities that are clad with gold mirrors. Mounted samples are held in place with a thin cover glass over the open end of the cavity

3.3.2 Cryostat

A liquid He immersion cryostat with quartz windows was used for this study. Most spectra were recorded at temperatures below the λ-point in superfluid He. Pressures as low as 3 mm Hg were achieved with a mechanical vacuum pump. The reduced He vapour pressure corresponds to a temperature of $\sim$1.5 K. In this study, the main advantage of working with superfluid helium is an improved signal-to-noise level as compared to helium at 4.2 K. In the superfluid state, helium boil-off bubbles disappear and do not interfere with the optical path. In addition, thermally activated excitations in the sample were reduced to a minimum. Higher sample temperatures of 4.2 K were used to observe higher energy components of the split IBE ground states.

Figure 3.8 shows how the samples were mounted in a brass sample holder. Slots were cut into the brass sheet to accommodate the samples, which were then covered with Au mirrors on two sides and a glass cover on the third side, through which the excitation and PL emission take place. These highly reflective cavities increased the PL yield. The samples were loosely held in the cavities, with their own weight being the only force on them to avoid any mechanical strain. Sufficient gaps were included in the sample holder to ensure He contact of the samples at all times.

3.3.3 Other Experimental Methods

The primary concern of this study is the observation of isotopic fingerprints of NP transition lines through PL spectroscopy. Other experiments, including absorption or PLE spectroscopy, have been done on some of the studied impurity centres in the past (see Chap. 2). These methods are not expected to significantly benefit from the use of ^{28}Si material and contribute to the study of isotopic fingerprints because they are used mainly to study the electronically excited state spectrum. Furthermore, the samples used here are usually very small, therefore they do not lend themselves to absorption experiments. Similar arguments can be made for stress and Zeeman spectroscopy, for which the studied PL lines often would not be bright enough.

References

1. M. Thewalt, M. Steger, A. Yang, M. Cardona, H. Riemann, N. Abrosimov, M. Churbanov, A. Gusev, A. Bulanov, I. Kovalev, A. Kaliteevskii, O. Godisov, P. Becker, H.-J. Pohl, Can highly enriched ^{28}Si reveal new things about old defects? Phys. B **401–402**, 587–592 (2007)
2. A. Yang, M. Steger, M. Thewalt, M. Cardona, H. Riemann, N. Abrosimov, M. Churbanov, A. Gusev, A. Bulanov, I. Kovalev, A. Kaliteevskii, O. Godisov, P. Becker, H.-J. Pohl, J. Ager III, E. Haller, High resolution photoluminescence of sulphur- and copper-related isoelectronic bound excitons in highly enriched ^{28}Si. Phys. B **401–402**, 593–596 (2007)
3. M. Steger, A. Yang, N. Stavrias, M. Thewalt, H. Riemann, N. Abrosimov, M. Churbanov, A. Gusev, A. Bulanov, I. Kovalev, A. Kaliteevskii, O. Godisov, P. Becker, H.-J. Pohl, Reduction of the linewidths of deep luminescence centers in ^{28}Si reveals fingerprints of the isotope constituents. Phys. Rev. Lett. **100**, 177402 (2008)
4. M. Steger, A. Yang, M. Thewalt, H. Riemann, N. Abrosimov, P. Becker, H.-J. Pohl, High resolution photoluminescence of copper, silver, gold and lithium-related isoelectronic bound excitons in highly enriched ^{28}Si, in *AIP Conference Proceedings*, vol. 1199, ed. by M. Caldas, N. Studart (American Institute of Physics, Melville, 2010), pp. 33–34
5. M. Steger, A. Yang, T. Sekiguchi, K. Saeedi, M. Thewalt, M. Henry, K. Johnston, H. Riemann, N. Abrosimov, M. Churbanov, A. Gusev, A. Bulanov, I. Kaliteevski, O. Godisov, P. Becker, H.-J. Pohl, Isotopic fingerprints of gold-containing luminescence centers in ^{28}Si. Phys. B **404**, 5050–5053 (2009)
6. M. Steger, A. Yang, T. Sekiguchi, K. Saeedi, M. Thewalt, M. Henry, K. Johnston, E. Alves, U. Wahl, H. Riemann, N. Abrosimov, M. Churbanov, A. Gusev, A. Kaliteevskii, O. Godisov, P. Becker, H.-J. Pohl, Isotopic fingerprints of Pt-containing luminescence centers in highly enriched ^{28}Si. Phys. Rev. B **81**, 235217 (2010)
7. P. Becker, D. Schiel, H.-J. Pohl, A. Kaliteevski, O. Godisov, M. Churbanov, G. Devyatykh, A. Gusev, A. Bulanov, S. Adamchik, V. Gavva, I. Kovalev, N. Abrosimov, B. Hallmann-Seiffert, H. Riemann, S. Valkiers, P. Taylor, P.D. Bièvre, E. Dianov, Large-scale production of highly enriched ^{28}Si for the precise determination of the Avogadro constant. Meas. Sci. Technol. **17**, 1854–1860 (2006)
8. P. Becker, H.-J. Pohl, H. Riemann, N. Abrosimov, Enrichment of silicon for a better kilogram. Phys. Stat. Sol. B **207**, 49–66 (2010)
9. B. Andreas, Y. Azuma, G. Bartl, P. Becker, H. Bettin, M. Borys, I. Busch, M. Gray, P. Fuchs, K. Fujii, H. Fujimoto, E. Kessler, M. Krumrey, U. Kuetgens, N. Kuramoto, G. Mana, P. Manson, E. Massa, S. Mizushima, A. Nicolaus, A. Picard, A. Pramann, O. Rienitz, D. Schiel, S. Valkiers, A. Waseda, Determination of the Avogadro constant by counting the atoms in a ^{28}Si crystal. Phys. Rev. Lett. **106**, 030801 (2011)
10. A. Yang, M. Steger, D. Karaiskaj, M. Thewalt, M. Cardona, K. Itoh, H. Riemann, N. Abrosimov, M. Churbanov, A. Gusev, A. Bulanov, A. Kaliteevskii, O. Godisov, P. Becker, H.-J. Pohl, J. Ager III, E. Haller, Optical detection and ionization of donors in specific electronic and nuclear spin states. Phys. Rev. Lett. **97**, 227401 (2006)
11. M.L.W. Thewalt, U.O. Ziemelis, P.R. Parsons, Enhancement of long lifetime lines in photoluminescence from Si:In. Solid State Commun. **39**, 27–30 (1981)
12. J. Weber, R. Sauer, P. Wagner, Photoluminescence from a thermally induced indium complex in silicon. J. Lumin. **24–25**, 155–158 (1981)

13. A.A. Istratov, E.R. Weber, Physics of copper in silicon. J. Electrochem. Soc. **149**, G21–G30 (2002)
14. M. Henry, S. Daly, C. Frehill, E. McGlynn, C. McDonagh, A photoluminescence study of gold- and platinum-related defects in silicon using radioactive transformations, in *Physics of Semiconductors: 23rd International Conference on the Physics of Semiconductors—ICPS 1996*, ed. by M. Scheffler, R. Zimmermann (World Scientific, Singapore, 1996), pp. 2713–2716
15. M.O. Henry, E. Alves, J. Bollmann, A. Burchard, M. Deicher, M. Fanciulli, D. Forkel-Wirth, M.H. Knopf, S. Lindner, R. Magerle, R. McGlynn, K.G. McGuigan, J.C. Soares, A. Stotzler, G. Weyer, Radioactive isotope identifications of Au and Pt photoluminescence centres in silicon. Phys. Stat. Sol. B **210**, 853 (1998)
16. M. Henry, M. Deicher, R. Magerle, E. McGlynn, A. Stotzler, Photoluminescence analysis of semiconductors using radioactive isotopes. Hyperf. Int. **129**, 443–460 (2000)
17. D. Forkel-Wirth, Radioactive ion beams in solid state physics. Phil. Trans. R. Soc. Lond. A **356**, 2137–2162 (1998)
18. C.A.J. Ammerlaan, A.B. van Oosten, Electronic structure of platinum in silicon. Phys. Scripta **25**, 342–347 (1989)
19. R.C. Jaeger, *Introduction to Microelectronic Fabrication*, 2nd edn. (Prentice Hall, Prentice, 2002)
20. J.F. Ziegler, *The Stopping and Range of Ions in Matter—SRIM*, http://www.srim.org, 2010
21. A. Michelson, Visibility of interference-fringes in the focus of a telescope. Phil. Mag. **31**, 256–259 (1891)
22. A. Michelson, On the application of interference methods to spectroscopic measurements. Phil. Mag. **34**, 280 (1892)
23. P.R. Griffiths, J.A. de Haseth, *Fourier Transform Infrared Spectroscopy* (Wiley, New York, 1986)
24. J. Kauppinen, D. Moffatt, D. Cameron, H. Mantsch, Noise in Fourier self-deconvolution. Appl. Opt. **20**, 1866 (1981)

Chapter 4
Results

In this chapter we review the results of isotopic fingerprints obtained using highly enriched ^{28}Si. Although some results have been published before [1–7], many new details are presented here. The subsections are organized in a partially historical way, following the developments as various, previously known and new centres were studied in ^{28}Si. This chapter is also organized in a way that anticipates our proposal that the four- and five-atom containing TM centres have, at their core, a substitutional TM, surrounded by interstitial TM, usually Cu, but also Ag or the non-TM Li. This model clearly does not apply to the S-containing S_A and S_B centres, which are included because they are IBE centres that also contain Cu.

4.1 Cu and Ag-Related Centres

In this section, defect centres related to Cu and Ag will be introduced, including those cases where Li is seen to substitute for Cu. Figure 4.1 shows an overview of the PL of these complexes. The energies of their NP lines are summarized in Table 4.1 together with the relative intensities of different electronic states. The isotope shifts of the NP lines per atomic mass unit (amu) are given in Table 4.2. All of these centres also have low energy pLVM, replicas which are shown in Fig. 4.2 on a scale displaying their shift from their respective NP line. The replica energy shift, FWHM, and, where available, the measured isotope shift expressed as the ratio of the pLVM energy for a sample doped predominantly with the heavier isotope of the relevant species divided by the pLVM energy in a sample doped predominantly with the lighter isotope are tabulated in Table 4.3. The observed relative isotope shifts of the pLVM are compared to the simplest possible model, in which the pLVM energy goes as the inverse square root of the mass. Note that these tables and figures also show data for Au-related complexes which are discussed in Sect. 4.2.

M. Steger, *Transition-Metal Defects in Silicon*, Springer Theses,
DOI: 10.1007/978-3-642-35079-5_4, © Springer-Verlag Berlin Heidelberg 2013

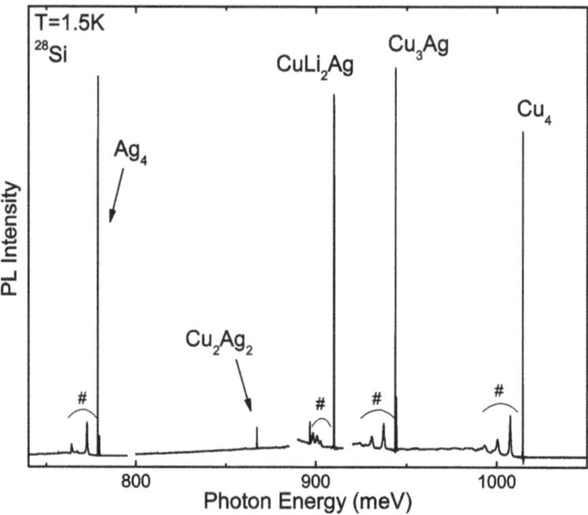

Fig. 4.1 Overview of the luminescence of complexes involving Cu, Ag, and Li at low spectral resolution. The unresolved NP lines are labelled with the identity of the complex, whereas the "#" marks some of the prominent pLVM replicas

Table 4.1 Energies (in meV) at 1.5 K of the NP transitions of all identified Cu, Ag, and Au containing complexes in ^{28}Si, including all states observable at liquid He temperatures

Complex	Energy	Intensity	Complex	Energy	Intensity
Cu$_3$Au {0}[†]	734.579	1	Cu$_2$Ag$_2$	867.389	
Cu$_3$Au {1}[†]	735.059	0.48	CuLi$_2$Ag	909.848	
Cu$_2$LiAu	735.177	0.04	Cu$_3$Ag {*Cu$_0^0$}[◇]	943.704	0.05
Cu$_2$LiAu	735.231	1	Cu$_3$Ag {*Cu$_0^0$}[◇]	943.753	1
CuLi$_2$Au	746.714	1	Cu$_4$ {Cu$_0^0\Gamma_4$}[††]	1014.435	1
CuLi$_2$Au	748.670	0.04	Cu$_4$ {Cu$_0^0\Gamma_3$}[††]	1014.576	0.14
Li$_3$Au	764.573	0.02	Cu$_3$LiAu	1052.744	
Li$_3$Au	764.821	0.23	Cu$_2$Li$_2$Au	1063.223	0.05
Li$_3$Au	764.906	0.03	Cu$_2$Li$_2$Au	1063.442	0.73
Li$_3$Au	765.274	1	Cu$_2$Li$_2$Au	1063.985	1
Li$_3$Au	765.393	0.26	Cu$_4$Au {FeB$_1^0$}[‡]	1066.649	
Ag$_4$ {F_0}[♯]	778.667	0.02	CuLi$_3$(Au)*	1090.199	
Ag$_4$ {A}[♯]	778.879	1			
Ag$_4$ {B}[♯]	779.767	0.07			

For those complexes containing Pt, see Table 4.4. The energies are given for the specific complexes where any Li is ^{6}Li, any Cu is ^{63}Cu, any Ag is ^{107}Ag and any Au is ^{197}Au. Errors are smaller than 2 μeV. The relative intensities for the different electronic transitions are measured from low resolution spectra at 4.2 K or adjusted to 4.2 K from lower temperature data, and normalized to 1 for the strongest electronic transition of each centre. The labels given in braces are assignments made in previous publications, with references to their first appearance: [†] Ref. [8]; [♯] Refs. [9, 10]; [◇] Ref. [11]; [††] Ref. [12]; [‡] Ref. [13]

Table 4.2 The observed isotope shifts of the NP lines of the centres studied here, in μeV per atomic mass unit (amu), relative to the energy for the complex where any Li is ^{6}Li, any Cu is ^{63}Cu, any Ag is ^{107}Ag, and any Au is ^{197}Au

Complex	Energy	Li	Cu	Ag	Au
Cu_3Au	734.579		1.9		2.3
Cu_2LiAu	735.231	33	3.2		2.0
$CuLi_2Au$	746.714	38	4.0		1.4
Li_3Au	764.821	33			$\sim$0
Ag_4 {A}	778.879			3.1	
Cu_2Ag_2	867.389		10	1.6	
$CuLi_2Ag$	909.848	145	20	2.2	
Cu_3Ag	943.753		11	1.5	
Cu_4 {Γ_3}	1014.576		9.5		
$Li_4\ V_{Si}$ "Q_L"[†]	1044	237			
Cu_3LiAu	1052.744	64	3.0		2.7
Cu_2Li_2Au	1063.985	35	4.0		2.6
Cu_4Au	1066.649		2.0		2.2
$CuLi_3(Au)^*$	1090.199	35	4.1		

For those centres containing Pt, see Table 4.5. The Au content of the centres labeled with "*" could not be verified with isotopic fingerprints, so no isotope shift value is available. [†] Values for the Li_4 V_{Si} "Q_L" centre, which was not studied here, are taken from Ref. [14]. PL energies are in meV

Fig. 4.2 Low resolution spectra of the pLVM replicas of all centres listed in Table 4.3 are shown versus the energy shift from the respective NP transitions. Labels for previously reported replicas are assigned according to the references given together with that table. The Cu_2LiAu phonon mode slightly overlaps with the Cu_3Au replicas, as indicated. Unidentified NP lines are marked with a "⋆"

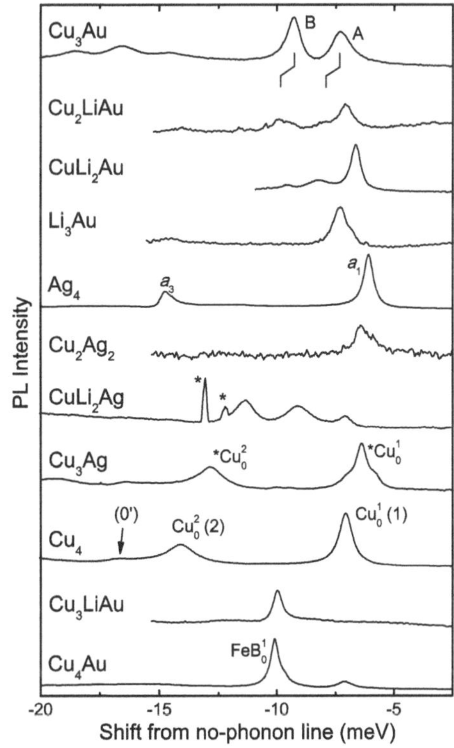

4.1.1 The 1014 meV Cu₄ Complex

As discussed in Sect. 2.1, the $\sim$1014 meV PL line was long thought to be due to a Cu-pair centre. Only recently, as part of this thesis, has it been shown conclusively that it is a four Cu-atom complex [1, 3]. Figure 4.3 shows the Cu_4 NP lines of the Γ_4 and Γ_3 electronic states [12] with their lowest energy line at 1014.44 and 1014.58 meV respectively. While the lower energy Γ_4 state shows a complicated splitting, the Γ_3 state is straightforward to analyze. The lines labeled with numbers from 0 to 4 stand for the number of ^{65}Cu in the Cu_4 complex, so the lowest energy line is due to complexes with 4 ^{63}Cu atoms. By replacing the lighter isotopes with ^{65}Cu one-by-one, we reach the highest energy line, which is due to 4 ^{65}Cu atoms. In the spectrum shown in Fig. 4.3 the relative intensities of those five lines reflect the natural abundance of the two stable Cu isotopes in a four-atom cluster very accurately. The extra structure seen for the Γ_4 transition was explained as being due to the effects of different arrangements of the two Cu isotopes in complexes containing mixed isotopes, because the 0 and 4 lines are seen to be unsplit [1, 3]. Whereas the overall structure of Γ_3 seems to be much simpler than that of Γ_4, a small splitting can be observed for line 1 of Γ_3, whereas lines 2 and 3 show a slight broadening. Tables 4.1 and 4.2 list the energy of the lowest energy line in ^{28}Si and the isotope shift per atomic mass unit (amu) of Cu_4, respectively. The experimental results have led to a renewed interest in modeling the Cu_4 centre, as discussed in Sect. 2.1.

In Fig. 4.2 a spectrum of the pLVM replicas of Cu_4 is shown on a scale that shows their shift from the NP line. The replica energy, its FWHM, and the measured isotope shift, in this case as the ratio $r_E = E_p(^{63}Cu)/E_p(^{65}Cu)$, are listed in Table 4.3. Our data for the isotope shift for Cu_4 of 1.013 compares well to the ratio of 1.010 measured by Weber et al. [12] and is slightly lower than the expected shift of 1.016, according to the simple model discussed above.

4.1.2 The 944 meV Cu₃Ag Complex

The Cu-containing centre at $\sim$944 meV was often referred to as $^{*}Cu$, as described in Sect. 2.2, and was thought to be related to the 1014 meV Cu_4 centre. The PL spectra [3] of this centre in ^{28}Si as a function of Cu isotopic composition are shown in Fig. 4.4, demonstrating that the centre contains three Cu atoms with a further splitting of almost equal intensity (labelled "a" and "b") that can be attributed to the stable Ag isotopes ^{107}Ag and ^{109}Ag, which have almost equal natural abundance.

The effects of Ag isotope substitutions for this centre are shown on the right side of Fig. 4.5, conclusively proving the incorporation of one Ag atom. Like Cu, Ag is a rapid interstitial diffuser in the positive charge state and comes from the same column of the periodic table. The Cu_3Ag centre has two electronic states with the lowest energy NP transitions (for ^{63}Cu and ^{107}Ag) at 943.70 and 943.75 meV, of which the higher one is stronger, as summarized in Table 4.1.

Table 4.3 Centres, excluding those containing Pt, with significant low energy pLVM replicas are listed with their respective pLVM energies E_p (in meV, errors are typically ± 0.02 meV), and FWHM (in meV)

	E_p	FWHM	$r_E(\mathrm{Cu})$	$r_E(\mathrm{Ag})$
Cu_3Au {A}[†]	7.35	0.81	*	
Cu_3Au {B}[†]	9.41	0.71	1.011(1)	
Cu_2LiAu	7.14	0.62	*	
$CuLi_2Au$	6.70	0.48	*	
$CuLi_2Au$	8.24	n.a.	n.a.	
$CuLi_2Au$	9.63	n.a.	n.a.	
Li_3Au	7.30	0.71	*	
Ag_4 {a_1}[♯]	5.91	0.51		1.008(2)
Ag_4 {a_3}[♯]	14.70	0.38		1.006(2)
Ag_4 {a_4}[♯]	22.12	0.31		1.010(2)
Cu_2Ag_2	6.48	0.76	n.a.	n.a.
$CuLi_2Ag$	7.23	0.90	n.a.	n.a.
$CuLi_2Ag$	9.20	1.14	n.a.	n.a.
$CuLi_2Ag$	11.51	1.43	n.a.	n.a.
Cu_3Ag {$^\star Cu_0^1$}[◊]	6.44	0.67	n.a.	1.006(1)
Cu_4 {Cu_0^1 (1)}[††]	7.03	0.69	1.013(1)	
Cu_3LiAu	10.04	0.46	n.a.	
Cu_4Au {FeB_1^1}[‡]	9.74	0.45	1.013(1)	

No higher order or combination mode replicas are listed, although they are often observed. Where available, the isotope shifts are given as the ratio r_E of the observed pLVM energy in a sample doped predominantly with the lighter isotope of the relevant species versus that in a sample doped with the heavier isotope. These ratios can be compared to the ratios r_m given in Table 1.1. No statistically significant isotope shift could be observed for replicas labeled "*"; for "n.a." no data was available. The labels given in braces are assignments made in previous publications, with references to their first appearance: [†] Ref. [8]; [♯] Ref. [9]; [◊] Ref. [11]; [††] Refs. [12, 15]; [‡] Ref. [13]

While no Cu isotope shift could be observed in the pLVM structure of Cu_3Ag, a Ag isotope shift of 1.006 was seen, which compares to an expected shift of 1.009. Further details can be seen in Fig. 4.2 and Table 4.3.

4.1.3 The 867 meV Cu_2Ag_2 Complex

The discovery of Cu_3Ag suggests that a series of centres of the form $Cu_xAg_{(4-x)}$ may exist, and a new centre was found in the same samples with its lowest energy component at 867.39 meV, shown on the left side of Fig. 4.5. The triplet structure with respect to both the Cu and Ag isotopes reveals it to be a Cu_2Ag_2 centre [3]. The components labeled "a" through "c" denote from 0 to 2 ^{109}Ag in the centre. While spectra of Cu_2Ag_2 for different Cu isotopic enrichments are not available, it is possible to compare the intensities of the different Cu components of the spectrum of Cu_2Ag_2 to those of Cu_3Ag (shown on the right side of Fig. 4.5) and Cu_4 in the

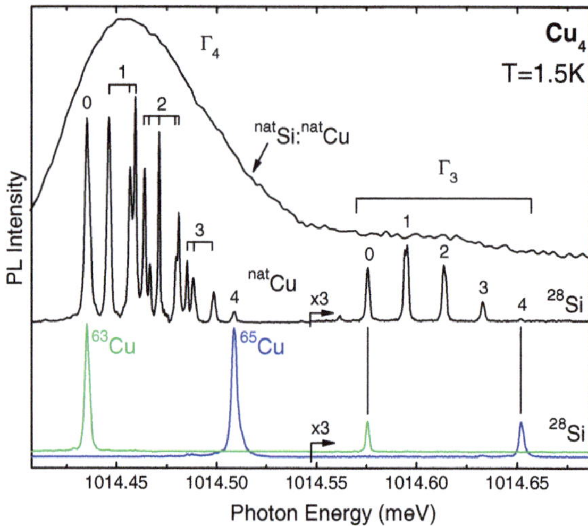

Fig. 4.3 The figure shows the Cu_4 NP lines ($\sim$1014 meV) in ^{nat}Si diffused with ^{nat}Cu (*top*, downshifted by 62 μeV for clarity). It is compared with those in ^{28}Si samples diffused with either ^{nat}Cu (*middle*), or enriched ^{63}Cu or ^{65}Cu (*bottom*). 0–4 denotes the number of ^{65}Cu in the Cu_4 complex. Γ_4 is the lowest energy ground state component, while Γ_3 is a component of the electronic ground state at slightly higher energy [12]. All spectra recorded at 1.5 K and resolutions better than 2.5 μeV

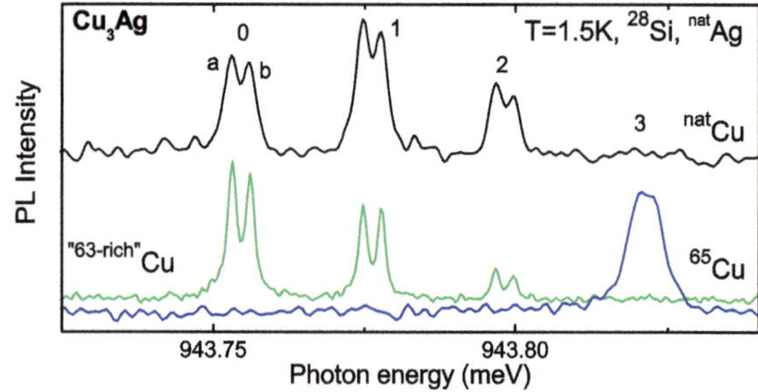

Fig. 4.4 The spectra show the NP luminescence of the Cu_3Ag centre in ^{28}Si for samples with different Cu isotopic compositions. Labels 0, 1, 2, and 3 give the number of ^{65}Cu contained in the centre

same sample. A consistent isotopic content of $\sim$44 % ^{65}Cu was found in all three cases. The relative intensities of the Ag components in the sample that contains ^{nat}Ag are in good agreement with the known natural Ag isotopic composition. Thus it is clear that the 867.39 meV PL centre consists of two Cu atoms in addition to two Ag atoms. The Cu_2Ag_2 PL has always been found to be much weaker than that of the Cu_3Ag PL, which, in samples containing Ag, can be comparable to or greater than the Cu_4 PL intensity.

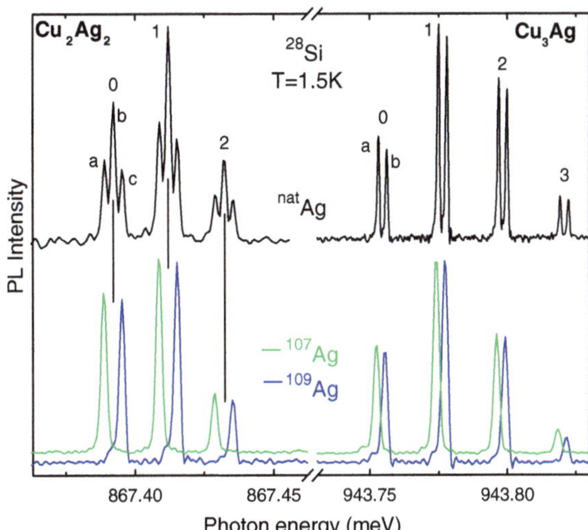

Fig. 4.5 The spectra show the NP lines of the Cu_3Ag and Cu_2Ag_2 centres in ^{28}Si containing Cu at near-natural abundance and either ^{nat}Ag, ^{107}Ag, or ^{109}Ag. Labels 0, 1, 2, and 3 give the number of ^{65}Cu contained in the centre, and "a, b, c" specify the number of ^{109}Ag, from 0 to 2

The PL of the Cu_2Ag_2 is very weak and, therefore only the energy and FWHM of the principal pLVM replica could be determined. Further details can be seen in Fig. 4.2 and Table 4.3.

While the sequence Cu_4–Cu_3Ag–Cu_2Ag_2 suggests that a $CuAg_3$ PL centre may exist, we were never able to find any evidence for such a centre, despite considerable effort.

4.1.4 The 778 meV Ag$_4$ Complex

Although no $CuAg_3$ centre was observed in our samples, it was found that the $\sim$778 meV centre, which has been previously attributed to a single Ag_s, indeed fits into this series as the Ag_4 complex [3]. This centre was always the most intense line in Ag-diffused ^{28}Si after quenching in the closed ampoule, but subsequent requenching, even with careful cleaning, shifted most of the PL intensity to the complexes containing both Ag and Cu. In Fig. 4.6 the top spectrum shows ^{28}Si diffused with ^{nat}Ag. Both the F_0 ground state with its lowest energy component "a" at 778.67 meV and the much stronger excited state A at 778.88 meV, as labeled by Davies et al. [10], split into five components, labeled "a" through "e" that denote from 0 to 4 ^{109}Ag in an Ag_4 centre, with their relative intensities accurately predicted by the ^{nat}Ag isotopic abundance. The spectra for the enriched ^{107}Ag and ^{109}Ag samples are dominated by the "a" and "e" components, respectively, as expected. Thus the 778 meV Ag centre was conclusively shown to be an Ag_4 complex [3], very similar to the Cu_4 centre. A third electronic state B was observed with its lowest energy fingerprint component

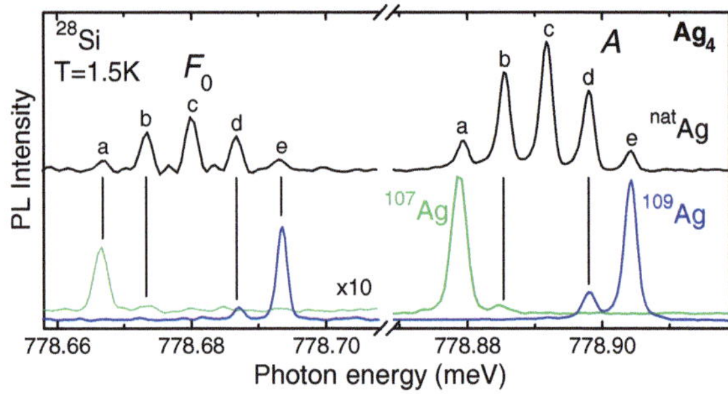

Fig. 4.6 The *upper* spectrum shows the NP lines of the $\sim$778.88 A transition, as well as the $\sim$778.67 meV F_0 transition of the Ag_4 complex for ^{nat}Ag in ^{28}Si. The lower spectra show the same transitions for samples diffused with either ^{107}Ag or ^{109}Ag. Labels "a" through "e" denote the number of ^{109}Ag, from 0 to 4 respectively, in the Ag_4 complex. All spectra recorded at 1.5 K and resolutions better than 2.5 μeV

at 779.77 meV. Tables 4.1 and 4.2 list the energies and isotope shifts per amu for the NP lines of the Ag_4 centre.

Several pLVM replicas could be observed for Ag_4, and Ag-related isotope shifts could be observed. The values are listed in Table 4.3 and are close to the expected shift of 1.009. Two of the pLVM replicas are shown in Fig. 4.2.

4.1.5 The CuLi₂Ag Complex

Following the discovery [4–6] that Li could replace Cu in the Au- and Pt-containing complexes, a sample that contained a mixture of the two stable Li isotopes, in addition to natural Cu and Ag, was produced. At first, this sample did not show any new Li-related PL centres, but similar to the Li-containing complexes of Au or Pt (see below), the previously unknown $CuLi_2Ag$ complex with its lowest energy line at 909.85 meV was observed after storing the sample at room temperature for an extended period. A spectrum of the NP line of this centre is shown in Fig. 4.7. Fits to the different components ("a" and "b", 0 and 1, respectively) confirm the content of ^{nat}Ag and ^{nat}Cu, whereas the relative intensities of the Li related components, labelled with letters K, L, M to designate the number of ^7Li atoms from 0 to 2, compare well to a different centre in a different sample with confirmed Li content that was diffused using the same Li source. No further Ag-Li centres could be observed, and no centres containing only Cu and Li were ever observed. Earlier reported Cu–Li centres [4] were discovered to also contain Au, as discussed in Sect. 4.2.4. Tables 4.1 and 4.2 list the energy and isotope shifts per amu for the $CuLi_2Ag$ centre. It is interesting to note that this complex shows remarkably large Li and Cu isotope shifts.

The $CuLi_2Ag$ centre has three strong pLVM replicas. Due to a shortage of ^{28}Si material, no further samples were produced to study possible isotope shifts of

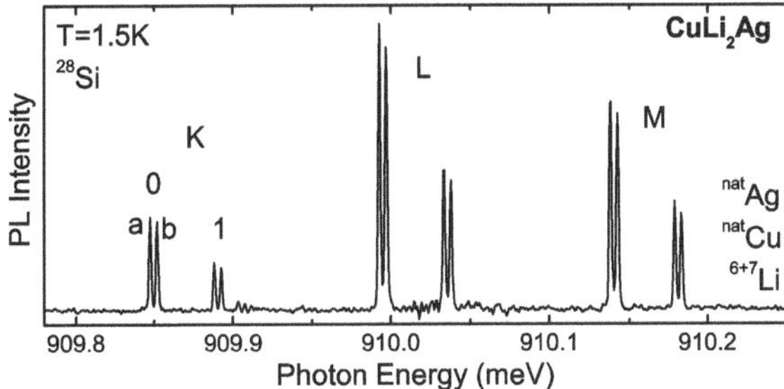

Fig. 4.7 The spectrum shows the NP lines of the ~909.85 meV CuLi$_2$Ag centre for a ^{28}Si sample containing ^{nat}Ag, ^{nat}Cu and a mixture of ^{6}Li and ^{7}Li. Labels "a" and "b" denote the number of ^{109}Ag, as 0 or 1 respectively; letters K, L, M designate the number of ^{7}Li atoms, from 0 to 2; and numbers 0 and 1 stand for the number of ^{65}Cu in the cluster. Spectrum recorded at 1.5 K and a resolution of 1.3 μeV

these modes. The energies of these replicas are listed in Table 4.3 and their spectrum is shown in Fig. 4.2. The sharp features at the low energy end of this spectrum are unidentified NP lines. No high energy LVM replicas that may result from Li motion were observed for this centre, but this might be explained by spectral interference from other PL centres in the region of interest.

4.2 Au-Related Centres

In this section, we discuss IBE PL centres that contain Au together with Cu, and/or Li. Figure 4.8 shows an overview of these complexes, and their NP energies are summarized in Table 4.1. The isotope shifts of the NP lines are given in Table 4.2. Many of these centres also have pLVM replicas that are shown in Fig. 4.2 and their energies are tabulated in Table 4.3.

4.2.1 The 735 meV Cu$_3$Au Complex

Like Ag, Au is in the same column of the periodic table as Cu and so it is another likely candidate for the formation of IBE complexes together with other TM and/or Li. Au is one of the most thoroughly studied deep metallic defects in Si. Au-diffused ^{28}Si samples have shown a dominant PL system with NP lines at ~735 meV [3], together with characteristic pLVM replicas, which have been observed previously in ^{nat}Si. As discussed in Sect. 2.4, this centre was originally attributed to Fe, whereas later studies showed the involvement of Au. As shown in the top panel in Fig. 4.9, in ^{28}Si the isotopic fingerprint of the 735 meV system reveals that it contains three Cu

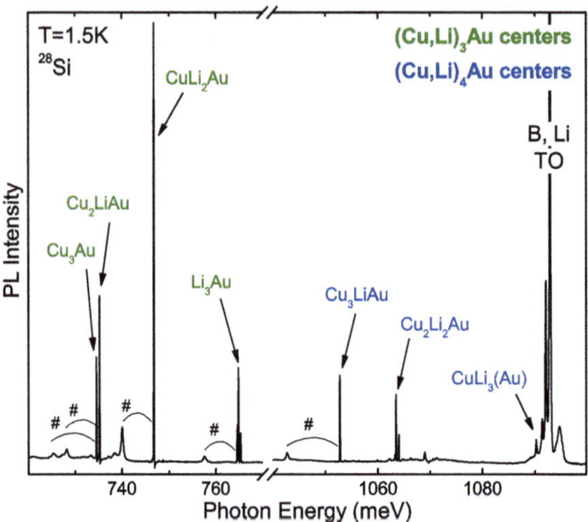

Fig. 4.8 Overview of the 4-atom $(Cu,Li)_3Au$ and the 5-atom $(Cu,Li)_4Au$ series at low spectral resolution. The unresolved NP transitions are labelled with the identity of the centre and "#" marks some of the prominent pLVM. At the high energy side are the transverse-optical (TO) phonon replicas of the shallow B and Li bound excitons

atoms [3], but until recently the number of Au atoms was unclear. This was remedied when ^{28}Si samples that were implanted with the unstable ^{195}Au radioisotope were used [5], allowing for the detection of an Au fingerprint in conjunction with the stable isotope ^{197}Au. Labels "α" and "β" were introduced to label PL lines due to a single ^{195}Au or a single ^{197}Au atom in the complex, respectively [5]. Only two peaks, "α" and "β", can be observed, indicating the presence of a single Au atom. The isotope shift for Au in this centre is larger than that for Cu, even though Au is much heavier. We can now conclusively say that the 735 meV system is a Cu_3Au complex.

The characteristic "A" and "B" pLVM replicas of the Cu_3Au centre are shown in Fig. 4.2. As seen in Fig. 4.10, the 7.35 meV "A" replica shows no significant shift with the change of Cu or Au isotopes, whereas the 9.41 meV "B" replica shows a significant Cu-shift. No significant Au shift was observed for either replica.

4.2.2 Four-Atom (Cu, Li)₃Au Complexes

Although initial suggestions [3] of a family of complexes with the members Cu_2Au_2 and $CuAu_3$ could not be confirmed, a different series of four-atom complexes that form in the presence of Li has been reported [5]. This $(Cu,Li)_3Au$ series is shown in Fig. 4.8 with high resolution details provided in Fig. 4.9.

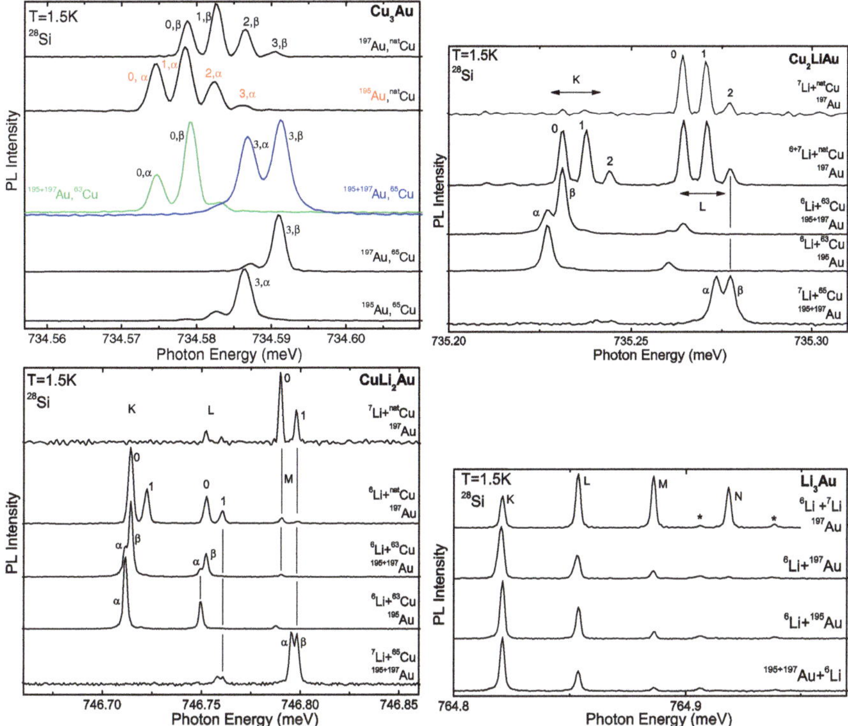

Fig. 4.9 Spectra of the 4-atom $Cu_xLi_{(3-x)}Au$ series in ^{28}Si. Numbers from 0 to 3 designate the number of ^{65}Cu atoms in the cluster; "α" labels a spectroscopic line due to ^{195}Au and "β" a line due to ^{197}Au; letters K, L, M, N designate the number of ^{7}Li atoms in the cluster, from 0 (K) to 3 (N). Peaks labeled "$\star$" are of a different electronic state. All spectra recorded at 1.5 K and resolutions better than $1.4\,\mu eV$

Li with natural isotope abundances as well as enriched ^{6}Li and ^{7}Li, was introduced into ^{28}Si samples, and the centres Cu_2LiAu, $CuLi_2Au$, and Li_3Au were found. Thus a series of four-atom $(Cu,Li)_3Au$ complexes is observed. In Fig. 4.9, these new family members are shown along with Cu_3Au in the order of increasing Li content. For Cu and Au, the familiar labels from 0 to 3, and "α" or "β", are used respectively to identify peaks due to different isotope combinations of Cu and Au, whereas labels K, L, M, and N are introduced to label the number of ^{7}Li atoms from 0 to 3 in each complex. In the case of Li_3Au, an Au isotope shift could not be determined reliably. In this series, a decreasing Au isotope shift is observed with an increasing amount of Li in the complex, so a smaller shift is expected for Li_3Au. As the Cu content reduces in the complexes, the Cu isotope shift increases while no definitive trend can be observed for Li. Tables 4.1 and 4.2 list the energies and isotope shifts per amu for each centre. For Cu_3Au, two different electronic states are observed, with their lowest energy PL lines at 734.58 and 735.06 meV, where the lower energy

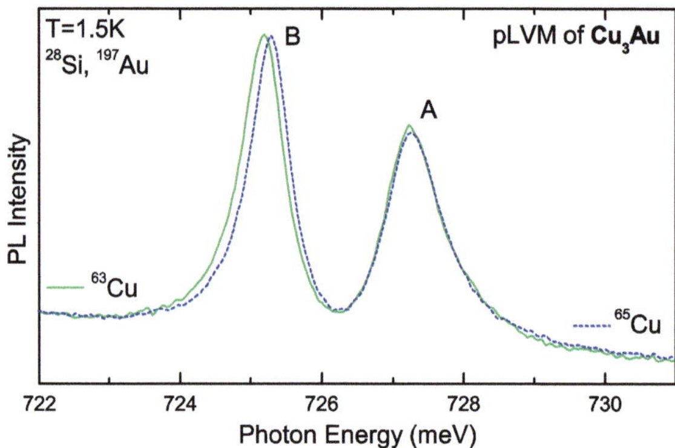

Fig. 4.10 Spectra of the pLVM replicas of the Cu$_3$Au centre in ^{28}Si diffused with ^{197}Au. The solid spectrum is due to ^{63}Cu; the dashed spectrum is due to ^{65}Cu. The "A" replica superimposes in both spectra, indicating that it has no Cu-related shift. The "B" replica shifts strongly with a change in Cu isotope composition. The ^{65}Cu spectrum has been downshifted by 11.5 μeV to compensate for the shift of the NP line. Both spectra recorded at 1.5 K and a resolution of 0.6 meV

state has the higher PL intensity. Cu$_2$LiAu has two states at 735.18 and 735.23 meV, with the higher one being much stronger. CuLi$_2$Au has electronic states at 746.71 and 748.67 meV, with the former being much stronger, and Li$_3$Au has 5 different electronic states at 764.57, 764.82 (strongest at 1.5 K), 764.91, 765.27 (strongest at 4.2 K), and 765.39 meV. For reasons of simplicity, Fig. 4.9 only shows transitions from one electronic state for each complex, although all states of a given centre showed similar fingerprints.

pLVM replicas were observed for all (Cu,Li)$_3$Au centres, as shown in Fig. 4.2 and summarized in Table 4.3. The relatively weak PL, together with overlap with other PL systems, prevented any isotope shift measurements for the pLVM replicas of the four-atom systems containing both Li and Au. We have observed a high energy LVM replica, presumably involving motion of Li, for only one Au-containing centre: CuLi$_2$Au. Its spectrum is shown in Fig. 4.19 and its energy is listed in Table 4.7, together with other high energy LVM replicas that were observed for the Pt-containing centres. Again, the absence of high energy LVM replicas for the other centres containing both Au and Li could simply be due to the weak PL of some of these systems, and the overlap of the relevant energy region with PL from other centres.

4.2.3 The 1066 meV Cu$_4$Au Complex

As reviewed in Sect. 2.5, the 1066.649 meV centre was originally thought to involve B and Fe, and it was often referred to as the Fe–B pair centre, while later the participation of Au was discovered. Only recently [5] the centre was shown to be Cu$_4$Au. The NP transition studied here is referred to as FeB$_1^0$ in previous studies [13]. The much

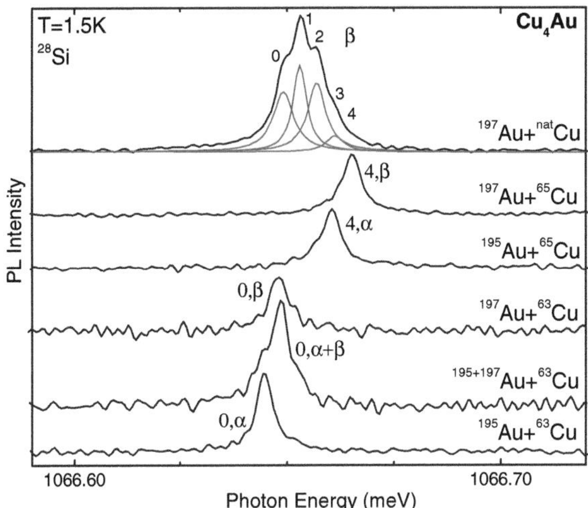

Fig. 4.11 Spectra of the Cu$_4$Au centre in ^{28}Si. Numbers from 0 to 4 designate the number of ^{65}Cu atoms in the cluster; "α" labels a spectroscopic line due to ^{195}Au and "β" a line due to ^{197}Au. The four peaks underneath the upper spectrum are the result of a fitting routine, omitting the weak peak 4. All spectra recorded at 1.5 K and resolutions of better than 2.5 μeV

weaker, lower energy FeB$_0^0$ NP PL could not be observed in our high resolution spectra with sufficient signal-to-noise ratio to be useful. Figure 4.11 shows that an isotopic fingerprint corresponding to four Cu atoms, as well as a single Au atom, can be observed. The top spectrum shows ^{nat}Cu with an underlying fit of four peaks representing peaks 0 to 3. Peak 4 due to four ^{65}Cu is too weak to be fit. In the lower spectra, enriched Cu was used together with ^{195}Au or ^{197}Au to produce single peaks. Figure 4.11 also shows one sample with mixed Au isotopes, in which the PL line appears slightly broader with a shoulder where the peak due to ^{195}Au is expected, which is of lower concentration in this sample than ^{197}Au. The signal to noise ratio is relatively small, but the Au isotope shift is clearly resolved, showing the "α" and "β" peaks. It should also be noted that the Cu and Au isotope shifts are close to those of Cu$_3$Au, but the width of the individual components is more than twice that of Cu$_3$Au. Therefore the individual peaks in a sample of mixed Cu or Au isotopes cannot be resolved in Cu$_4$Au. A sample diffused with ^{54}Fe was prepared to test whether Fe may be a part of the centre, but while the 1066.649 meV line was present, no Fe-related shift could be observed. Natural Fe has an abundance of 92.8 % ^{56}Fe, so a shift similar to the shift due to Cu isotopes should have been observable if the centre contained Fe.

The pLVM replicas of the Cu$_4$Au centre are dominated by the $\sim$9.7 meV mode previously reported by Sauer and Webar [13], which replicates both the FeB$_0^0$ and FeB$_1^1$ electronic transitions. A Cu isotope shift could be observed for this mode, but not an Au shift, as summarized in Table 4.3 and Fig. 4.2.

4.2.4 Five-Atom (Cu, Li)$_4$Au Complexes

The five atoms detected in the core of the Cu_4Au complex were a surprise at first, given the dominance of four-atom complexes detected by the isotopic fingerprint method until then, but the Cu_4Au complex was soon [5] found to be the prototype of a series of five-atom complexes with Li sequentially replacing Cu. A preliminary report [4] on these complexes prior to the studies [5] with the radioisotope ^{195}Au did not recognize the involvement of Au in those centres, therefore, at that time they were incorrectly labeled Cu_3Li, Cu_2Li_2, and $CuLi_3$, even though they were only observed in samples known to contain Au. The isotopic fingerprint of Au now clearly reveals an Au isotope shift for two of these complexes, as shown in Fig. 4.12. This figure shows in the upper two panels that Cu and Li isotope shifts, as well as an Au isotope shift due to a single Au atom, can be observed for Cu_3LiAu and Cu_2Li_2Au. The labeling scheme in the figure follows the usual pattern, with numbers 0 to 3, letters K, L, M, N, and Greek letters "α" and "β" standing for the number of ^{65}Cu, ^{7}Li, and ^{197}Au, respectively. The lower panel in Fig. 4.12 shows the $CuLi_3$(Au) centre for which the involvement of one Cu and three Li-atoms can be verified. The PL of this centre has always been weak and it is obscured by a nearby transverse-optical phonon replica of shallow bound exciton PL. It was only observed in Au-diffused rather than implanted samples, presumably due to the low implantation dose. Therefore the Au content of the centre cannot be verified by an isotopic fingerprint, but the evidence strongly suggests that it is part of the (Cu,Li)$_4$Au series. Unlike the trend for the four-atom (Cu,Li)$_3$Au complexes, for the five-atom complexes the Au isotope shift remains almost constant going from Cu_3LiAu to Cu_2Li_2Au, and the Li isotope shift decreases with increasing Li content, whereas the Cu isotope shift increases with decreasing Cu in the complex, similar to the behaviour seen for the four-atom complexes. As for the four-atom Au complexes, an increased Li content leads to a lower binding energy of the IBE. Although only one electronic state each could be observed for Cu_3LiAu (1052.75 meV) and $CuLi_3$(Au) (1090.20 meV), Cu_2Li_2Au shows three electronic states at 1063.22, 1063.44 meV (strongest at 1.5 K) and 1063.98 meV (strongest at 4.2 K). Tables 4.1 and 4.2 list the NP energies and isotope shifts per amu for this series.

Of these three complexes, only Cu_3LiAu shows a pLVM replica, but no significant isotope shift to the pLVM replica could be observed (see Table 4.3 and Fig. 4.2). The absence of pLVM replicas for the other two complexes likely results from the low PL intensity and interference from other PL systems in the region of interest.

4.3 Pt-Related Centres

As reviewed in Sect. 2.6, two of the deep PL centres discussed here were previously associated with Pt, namely the 777 and 884 meV centres [16–18]. We first discuss these two centres and then focus on other newly discovered Pt complexes that are

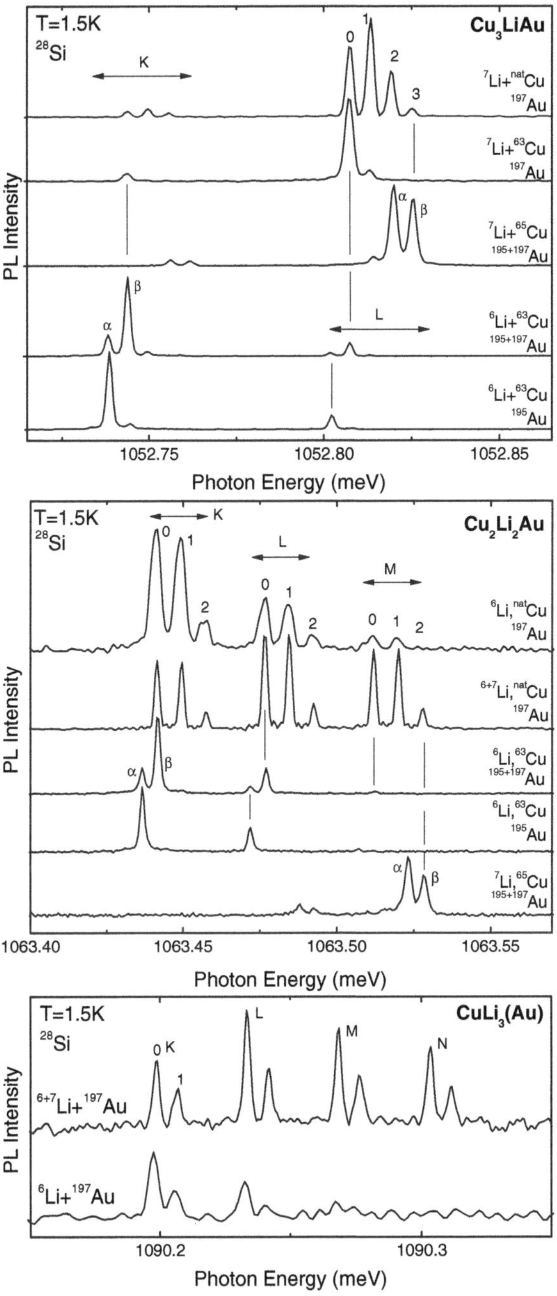

Fig. 4.12 Spectra of the 5-atom $Cu_xLi_{(4-x)}Au$ series in ^{28}Si. Numbers from 0–4 designate the number of ^{65}Cu atoms in the cluster; "α" labels a spectroscopic line due to ^{195}Au and "β" a line due to ^{197}Au; letters K, L, M, N designate the number of ^7Li atoms in the cluster, from 0 (K) to 3 (N). All spectra recorded at 1.5 K and resolutions better than $1.4\,\mu eV$

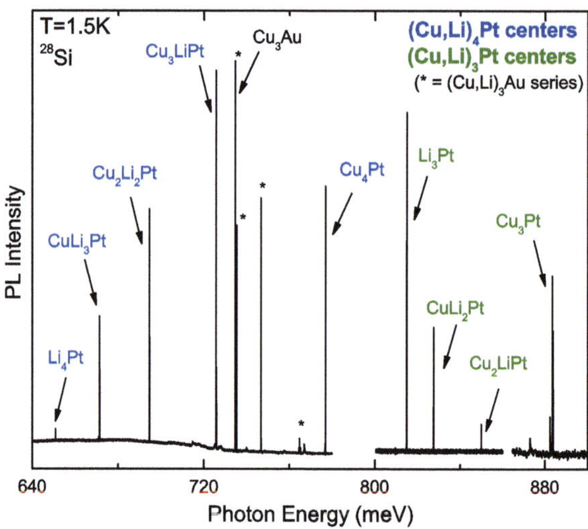

Fig. 4.13 Overview of the 4-atom $(Cu,Li)_3Pt$ and the 5-atom $(Cu,Li)_4Pt$ series taken at high resolution, and thus showing only the NP transitions. Peaks labeled with a "★" are Au-related centres (the ^{195}Pt resulted from the decay of ^{195}Au, some of which remains in the sample [5])

related to these centres. When isotopic fingerprints for the 777 and 884 meV centres were first reported in samples containing ^{195}Pt [5], the number of Pt atoms could not be determined. Only a further study involving samples diffused with other stable Pt isotopes proved the involvement of a single Pt atom [6]. The previously reported 1026 meV Pt-related centre [18, 19] was not observed in any of our Pt-containing ^{28}Si samples.

Figure 4.13 shows an overview of all of the NP lines of the Pt-containing deep centres discussed here, including the 777 meV Cu_4Pt complex, the 884 meV Cu_3Pt complex, and the four-atom $(Cu,Li)_3Pt$ and the five-atom $(Cu,Li)_4Pt$ series resulting from Li substituting for Cu [6]. Note that these spectra were taken at high resolution and therefore the relatively broad pLVM and LVM replicas are not seen, unlike in Fig. 4.1. Many of the associated LVM and pLVM replicas are detailed in the lower resolution spectra of Figs. 4.14 and 4.19.

The energies of the NP lines for all of the observed electronic states of the Pt-containing complexes reported here are summarized in Table 4.4 together with their relative intensities. Similarly, the NP isotope shifts per amu are listed in Table 4.5.

The limited availability of Pt samples and the weakness of some of the pLVM replicas, together with the small shifts expected for Pt, and the multiple Pt isotopes present in all but the ^{195}Pt samples, made it impossible to obtain Pt isotope dependent pLVM shifts for the Pt-containing centres. Nevertheless, a number of pLVM replicas could be observed and their energies are listed in Table 4.6. All observed Pt low energy pLVM replicas are shown in Fig. 4.14, where they are stacked on a scale showing the energy shift from their respective NP lines.

Table 4.4 Energies (in meV) at 1.5 K of all identified Pt-related complexes, including all states observable at liquid He temperatures

Complex	Energy (meV)	Intensity	Complex	Energy (meV)	Intensity
Li_4Pt	650.918	1	Li_3Pt	814.782	0.03
Li_4Pt	651.056	0.05	Li_3Pt	814.882	1
$CuLi_3Pt$	671.291	0.01	$CuLi_2Pt$	827.451	0.06
$CuLi_3Pt$	671.643	1	$CuLi_2Pt$	827.569	1
$CuLi_3Pt$	671.647	0.03	Cu_2LiPt	849.962	0.08
Cu_2Li_2Pt	694.571	0.20	Cu_2LiPt	850.131	1
Cu_2Li_2Pt	694.613	1	Cu_2LiPt	851.192	0.18
Cu_3LiPt	725.599	1	Cu_3Pt	882.360	0.22
Cu_3LiPt	725.829	0.13	Cu_3Pt	883.432	1
Cu_4Pt	776.779	0.01	Cu_3Pt	883.792	0.86
$Cu_4Pt\ \{A\}^{\dagger}$	776.926	1			

The energies are given specifically for centres where any Li is ^6Li, any Cu is ^{63}Cu, and any Pt is ^{195}Pt. The relative intensities for the different electronic transitions are extracted from low resolution files at 4.2 K. Errors are smaller than $\pm 2\,\mu eV$. [†]Label assigned in Ref. [16]

Table 4.5 The observed Li, Cu, and Pt isotope shifts of the NP lines of the centres studied here, in μeV per atomic mass unit (amu), as compared to the energy of the transition of the specific complex where any Li is ^6Li, any Cu is ^{63}Cu, and any Pt is ^{195}Pt

Complex	Energy	Li	Cu	Pt
Li_4Pt	650.918	41		3.1
$CuLi_3Pt$	671.643	46	4.1	3.6
Cu_2Li_2Pt	694.571	51	4.3	2.8
Cu_3LiPt	725.599	86	2.9	2.8
Cu_4Pt	776.779		1.7	2.1
Li_3Pt	814.882	52		4.2
$CuLi_2Pt$	827.569	50	5.8	3.0
Cu_2LiPt	850.131	70	2.3	2.9
Cu_3Pt	882.360		3.6	2.2

PL energies are in meV

4.3.1 The 777 meV Cu_4Pt Complex

Samples containing several different Pt isotopes were produced, and the number of Pt atoms in all centres shown here was determined to be one [6], similar to the Au complexes introduced earlier [5]. Figure 4.15 shows the 777 meV centre in ^{28}Si for three samples with different Cu isotope content, demonstrating the involvement of four Cu atoms. To identify the different components of the Pt PL centres, the now familiar labeling scheme is used. In Fig. 4.15 the lines are labeled with numbers from 0 to 4, where the lowest energy line is due to complexes with 4 ^{63}Cu atoms and the highest energy line is due to a complex having 4 ^{65}Cu atoms. In the upper spectrum shown in Fig. 4.15, the relative intensities of these five lines accurately reflect the natural abundance of the two stable Cu isotopes in a centre containing four Cu

Table 4.6 Pt centres with significant low energy pLVM replicas are listed with their respective pLVM energies E_p (in meV, errors are typically ± 0.02 meV), and FWHM (in meV)

	E_p	FWHM	r_E(Cu)
Cu_2Li_2Pt	10.03	0.4	n.a.
Cu_3LiPt	10.73	0.5	n.a.
Cu_3LiPt	21.93	0.5	n.a.
Cu_4Pt (a)	9.82	0.3	1.012(1)
Cu_4Pt (2a)	19.21	0.6	1.012(3)
Cu_4Pt (b)	20.92	0.4	1.015(2)
Cu_4Pt (3a)	28.84	0.7	1.013(2)
Cu_4Pt (a+b)	30.75	0.5	1.015(2)
Cu_2LiPt	9.93	1.4	n.a.
Cu_3Pt	9.36	0.3	1.011(2)

Where available, the isotope shifts are given as the ratio r_E of the observed pLVM energy in a sample doped predominantly with ^{63}Cu versus that in a sample doped with ^{65}Cu. These ratios can be compared to the ratios r_m given in Table 1.1. For Cu_4Pt, labels "a" and "b" indicate different combination modes. For "n.a.", no data was available. No Pt-related isotope shifts could be observed

atoms. The two spectra with enriched Cu clarify the labeling scheme. The second spectrum from the bottom shows a sample containing only ^{63}Cu isotopes together with the Pt isotopes ^{194}Pt, ^{195}Pt, and ^{198}Pt, whereas the bottom most spectrum is obtained from a sample diffused with ^{nat}Pt and ^{63}Cu. This results in a pattern of PL lines that reflect the relative concentrations of the different Pt isotopes. While the implanted sample displays a Pt isotope pattern characteristic of this implantation, which is seen repeatedly for all other Pt-containing PL centres discussed here, the spectrum of the diffused sample shown at the bottom of Fig. 4.15 closely follows the natural abundances of Pt isotopes, with observed relative intensities of 0.337, 0.336, 0.252, and 0.075 for ^{194}Pt, ^{195}Pt, ^{196}Pt, and ^{198}Pt. Taken together, it is clear that only a single Pt atom is involved in the formation of this centre. Therefore the label Cu_4Pt was assigned [6] to this complex, which has two electronic states at 776.78 and 776.93 meV. It is clearly distinguishable from the nearby Ag_4 complex at 778.88 meV [3]. Table 4.4 shows the energies of the two electronic states of Cu_4Pt for the isotopes ^{63}Cu and ^{195}Pt. The energies of all other peaks can be determined with the isotope shifts per amu given in Table 4.5. The much stronger PL from the higher energy state of Cu_4Pt at 776.93 meV (not shown) is not as well resolved as that of the weaker, lower energy transition, which is, therefore, shown in Fig. 4.15.

Cu_4Pt has a complex replica structure of five pLVM replicas that are a result of combinations of two different modes. These can be seen in Fig. 4.14, where the first three replicas of Cu_4Pt are shown together with the pLVM replicas of other Pt containing centres. The energies of the Cu_4Pt pLVM are summarized in Table 4.6. It should be noted that the greater number of pLVM replicas reported for the Cu_4Pt centre likely does not reflect any fundamental difference from the other Pt-containing centres, but instead results from the much larger PL intensity of this complex, resulting in strong pLVM replicas and little interference from other PL centres.

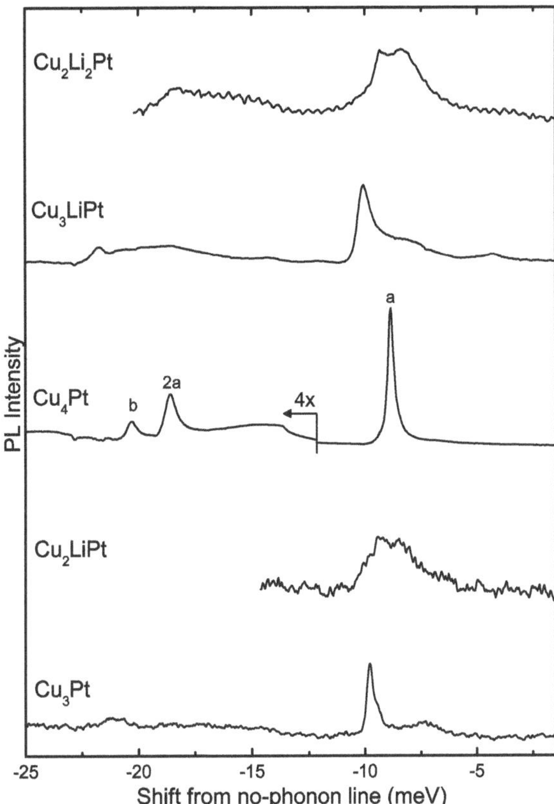

Fig. 4.14 Low resolution spectra of the pLVM replicas of all centres listed in Table 4.6 shown versus the energy shift from the NP line

4.3.2 The 884 meV Cu₃Pt Complex

The 884 meV Cu_3Pt system is shown in Fig. 4.16. Instead of five individual lines, as seen for Cu_4Pt in the $^{195}Pt+^{nat}Cu$ spectrum, this centre has only four, labeled 0–3, showing the number of Cu atoms in the complex to be three [6]. The two spectra with enriched Cu clarify this labeling scheme. The two spectra at the bottom show the pattern that is characteristic for our Pt implanted and diffused samples. The natural Pt abundance is visible in the lower spectrum, where the three lower energy peaks closely follow the expected relative intensities of ^{nat}Pt. A residual amount of ^{65}Cu remained visible in the ^{63}Cu diffused sample, resulting in a slightly obscured spectrum, due to the overlap of the $^{nat}Pt^{63}Cu$ ("0") components with the weaker $^{nat}Pt^{63}Cu_2{}^{65}Cu$ ("1") components. It was concluded that only one Pt atom is contained in this complex, which was therefore labeled Cu_3Pt [6]. The centre has three observable electronic states at 882.36 meV (strongest at 1.5 K), 883.43 meV

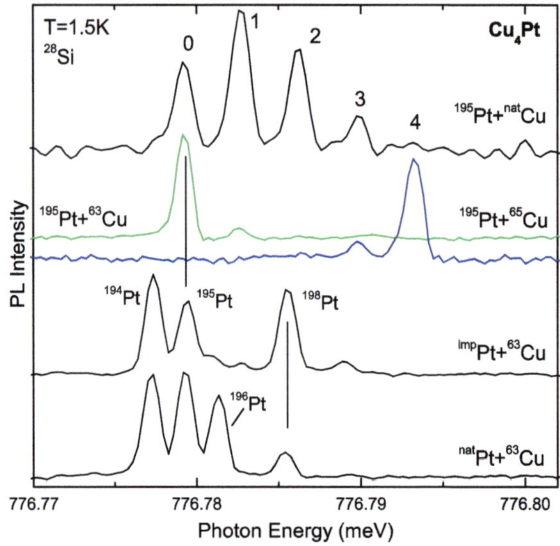

Fig. 4.15 NP transitions of the 776.78 meV component of the 777 meV Cu_4Pt system. The upper spectrum shows a sample diffused with ^{nat}Cu. Labels 0–4 denote the number of ^{65}Cu in the complex. Below this, two spectra show samples with enriched Cu. The *bottom* two spectra show samples that have been either implanted with Pt isotopes or diffused with natural Pt. The relative intensities of the PL lines in the ^{nat}Pt spectrum mirrors the isotopic abundances of natural Pt. The ^{imp}Pt spectrum is slightly obscured by the presence of ^{65}Cu in the sample. All spectra recorded at 1.5 K and resolutions better than $1.4\,\mu eV$

(strongest at 4.2 K), and 883.79 meV. These energies and the isotope shifts of the three observed Cu_3Pt electronic states are listed in Tables 4.4 and 4.5.

4.3.3 Four-Atom (Cu, Li)₃Pt Complexes

We now come to the recently discovered four-atom complexes related to Cu_3Pt, where Li substitutes for Cu [6]. Figure 4.17 shows the three centres—Li_3Pt, $CuLi_2Pt$, and Cu_2LiPt. For all of them, the spectrum of the Pt implanted sample shows the characteristic pattern already discussed for Cu_4Pt and Cu_3Pt. A spectrum of a sample with ^{nat}Pt isotope abundances is only available for Li_3Pt due to the weakness of the PL in some of the samples. The relative intensities of 0.34, 0.35, 0.21, and 0.05 of the PL lines of this spectrum accurately follow the natural isotope abundances of ^{194}Pt, ^{195}Pt, ^{196}Pt, and ^{198}Pt respectively. Together, this is evidence that there is only a single Pt atom in all four-atom Pt centres. The figure also shows the isotope shifts caused by different Cu and Li isotopes. Table 4.4 lists the energies of these centres and their different electronic states, whereas Table 4.5 gives the isotope shifts per amu.

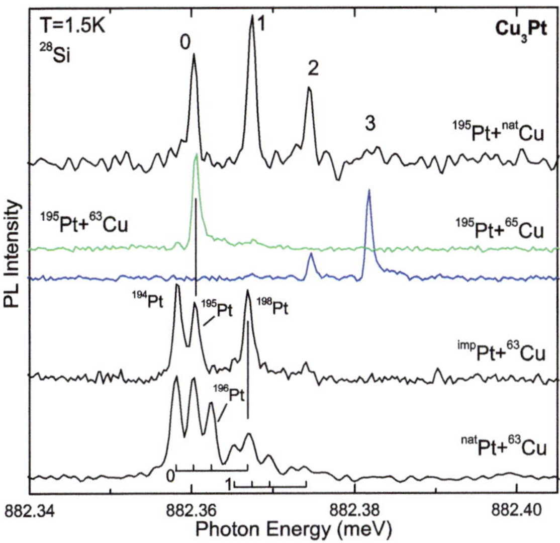

Fig. 4.16 NP transitions of the 884 meV Cu$_3$Pt system. The upper spectrum shows a sample diffused with ^{nat}Cu. Labels 0–3 denote the number of ^{65}Cu in the complex. Below this, two spectra show samples with enriched Cu. The bottom two spectra show samples that have been either implanted with Pt isotopes or diffused with natural Pt. The ^{nat}Pt+^{63}Cu sample also contains a small amount of ^{65}Cu, slightly obscuring the spectrum. All spectra recorded at 1.5 K and resolutions better than 1.4 μeV

PL of the four-atom Pt centre series and their pLVM replicas are, in general, relatively weak, therefore only replicas for Cu$_2$LiPt and Cu$_3$Pt could be observed. In Fig. 4.14, spectra of the low energy pLVM replicas are arranged on a scale showing their shift from the NP line. The energies and the Cu-isotope shift for Cu$_3$Pt are summarized in Table 4.6.

4.3.4 Five-Atom (Cu, Li)$_4$Pt Complexes

As was observed for Au-related complexes [5], Pt also forms five-atom complexes together with Cu and Li [6]. Figure 4.18 shows high resolution spectra of the NP transitions of these PL centres. For all four centres—Li$_4$Pt, CuLi$_3$Pt, Cu$_2$Li$_2$Pt, and Cu$_3$LiPt—a spectrum showing the characteristic Pt isotope pattern of our implanted sample is given and in addition to that, a spectrum of the ^{nat}Pt diffused sample is given for Li$_4$Pt. The relative intensities of 0.34, 0.32, 0.25, and 0.09 of the PL lines of this centre accurately follow the natural isotope abundances of ^{194}Pt, ^{195}Pt, ^{196}Pt, and ^{198}Pt, respectively. Therefore it is clear that a single Pt atom is part of each of these centres [6]. Additionally, spectra are displayed for each centre to illustrate

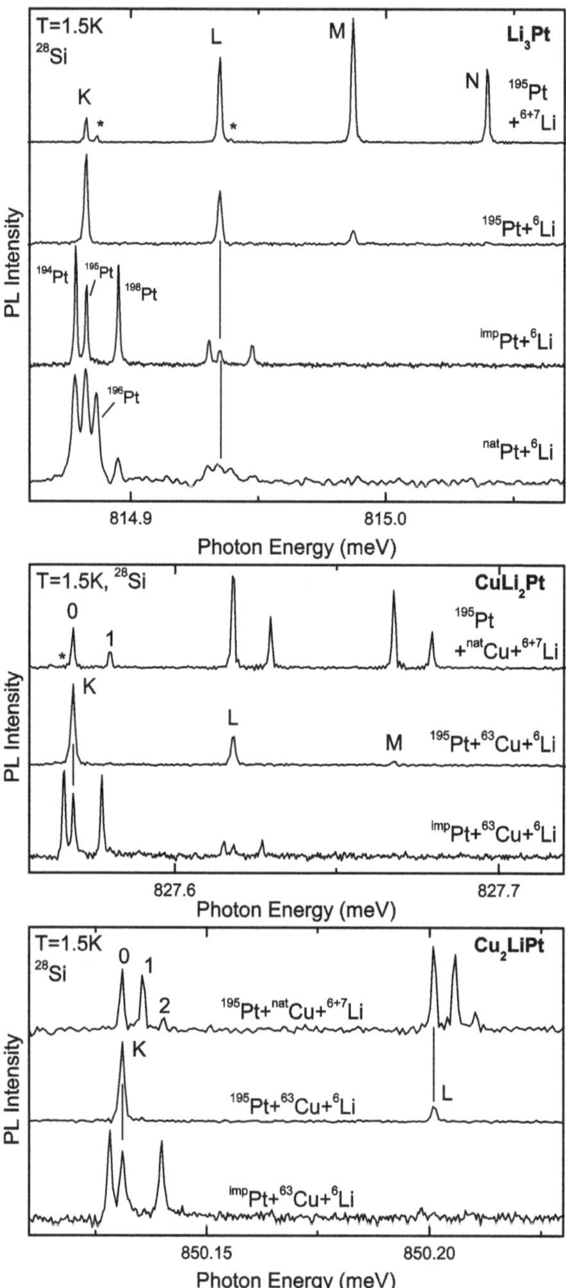

Fig. 4.17 Spectra of the NP transitions of the 4-atom $Cu_xLi_{(3-x)}Pt$ series in ^{28}Si. Numbers from 0 to 3 designate the number of ^{65}Cu atoms in the cluster, and letters K, L, M, and N designate the number of ^{7}Li atoms in the cluster from 0 (K) to 3 (N). Peaks labeled with a "⋆" are of a different electronic state. All spectra recorded at 1.5 K and resolutions better than 1.4 μeV

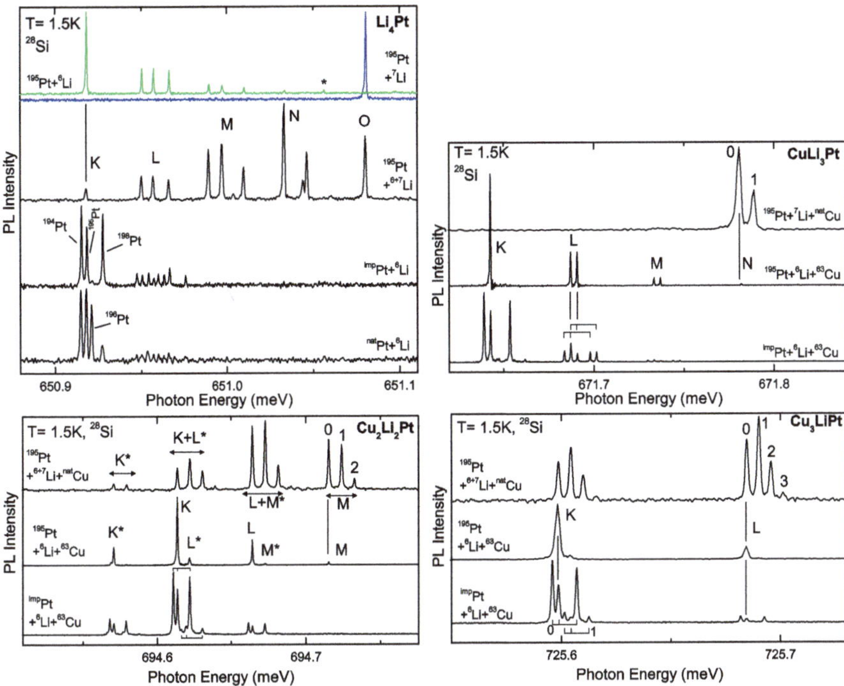

Fig. 4.18 Spectra of the NP transitions of the 5-atom $Cu_xLi_{(4-x)}Pt$ series in ^{28}Si. Numbers from 0 to 4 designate the number of ^{65}Cu atoms in the cluster, and letters K, L, M, N, and O designate the number of ^7Li atoms in the cluster from 0 (K) to 4 (O). Peaks labelled with a "$\star$" are from a different electronic state. All spectra recorded at 1.5 K and resolutions better than 1.4 μeV

how Cu is successively replaced by Li from Cu_4Pt to Li_4Pt (see Fig. 4.15), following the scheme $Cu_xLi_{(4-x)}Pt$. It should be noted that the Li_4Pt and $CuLi_3Pt$ centres displayed in Fig. 4.18 show an additional substructure for the complexes having mixed Li isotopes, which was not observed for other electronic states of these two centres. Similar substructure was observed before for the Γ_4 state of Cu_4 complexes having mixed Cu isotopes [3]. This structure is likely due to the fact that, in the present case, the Li sites of the complex are inequivalent, so that different arrangements of the Li isotopes result in slightly different PL energies. Table 4.4 lists the energies of these centres and their different electronic states; the isotope shifts per amu are found in Table 4.5.

Of the five-atom centres, the Cu_2Li_2Pt, Cu_3LiPt, and the Cu_4Pt centres had observable low energy pLVM replicas. The results are summarized in Table 4.6, where the mode energies are listed together with their full width at half maximum. Spectra of these replicas can be seen in Fig. 4.14.

The observed higher energy LVM replicas are shown in Fig. 4.19 and a list of energies of those modes is given in Table 4.7. A strong Li isotope shift of the high energy replicas can be seen, which is not surprising in that such high energy LVM

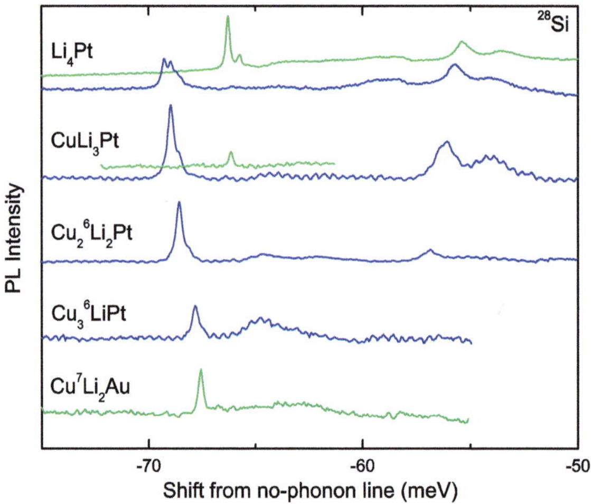

Fig. 4.19 Low resolution spectra of the Li-related high energy local vibrational mode replicas of all centres listed in Table 4.7. For the Li$_4$Pt and CuLi$_3$Pt centres, two spectra each are shown, whereas the respective *lower* one is from a ^{6}Li sample, while the *upper* one is from a ^{7}Li sample

Table 4.7 Four Pt centres and one Au centre with high energy Li local vibrational mode replicas are listed with the respective mode energies (in meV, errors are typically ± 0.02 meV) and FWHM (meV) Wherever available the replica energies are given for samples with ^{6}Li and ^{7}Li

	E	FWHM
^{6}Li$_4$Pt	69.30	0.3
^{7}Li$_4$Pt	66.44	0.2
Cu^{6}Li$_3$Pt	68.57	0.3
Cu^{7}Li$_3$Pt	66.00	0.2
Cu$_2^6$Li$_2$Pt	67.84	0.3
Cu$_2^7$Li$_2$Pt	65.50	0.2
Cu$_3^6$LiPt	67.09	0.4
Cu^{7}Li$_2$Au	65.12	0.2

usually involve the motion of a light atom. Such high energy LVM replicas lying between ~65 and ~70 meV have not been reported for the Li$_4$ V$_{Si}$ centre, as already noted in Sect. 2.8. The upper three spectra in Fig. 4.19 also show a feature at around -55 meV shift from the NP line, which seems reminiscent of the *B* replica of the Li$_4$ V$_{Si}$ centre, which is described as the symmetric breathing mode of the centre [14]. We have observed these high energy LVM replicas for only one Au-containing centre: the CuLi$_2$Au centre. Its spectrum is shown in Fig. 4.19 as well.

4.4 S-Related Centres

Although these IBE PL centres may not be directly related to the previously discussed centres, isotopic fingerprint studies have revealed that they do contain Cu, so the results are reviewed here for completeness. Two different configurations of the S-related centre are known [20]—S_A with NP PL at 968.2 meV and S_B at 812.0 meV. The two configurations show a metastability, with a conversion taking place from S_A to S_B under optical excitation at low temperature (T < 40 K) and a reverse conversion from S_B to S_A when heated to temperatures above 40 K. Thus observing the S_A centre at the low temperature needed for high resolution PL spectroscopy is difficult due to its rapidly diminishing intensity. The only ^{28}Si result for the S_A centre revealed a S isotope shift of 8 μeV for S_A as shown in Fig. 4.20.

4.4.1 The 812 meV S_B Complex

As reviewed in Sect. 2.7, it was suggested that the S_B centre consists of a substitutional S atom paired with an Cu_i, and S and Cu isotope effects were indeed observed in ^{28}Si [2]. The high resolution NP PL of S_B is shown in the two upper left spectra in Fig. 4.21, displaying a complicated substructure. The upper two spectra show an S isotope shift of 11 μeV between a sample containing ^{nat}S (95 % ^{32}S) and a ^{34}S diffused sample. The bottom spectra in this figure are shown along with the Γ_3 PL of the Cu_4 centre to demonstrate the actual Cu isotope content in each of the four different samples. At the bottom, spectra with enriched ^{63}Cu respectively ^{65}Cu are shown, which can be seen clearly in the Cu_4 spectrum. The actual isotopic enrichment was found to be ∼94 and ∼97 % for the ^{63}Cu and ^{65}Cu samples respectively. For S_B, each spectrum exhibits a peak with a substructure, which is unfortunately broadened in the case of ^{63}Cu. The labels "0" and "3" are attached to these lines, for 0, respectively 3 ^{65}Cu atoms in the complex. On either side of the triplet "3",

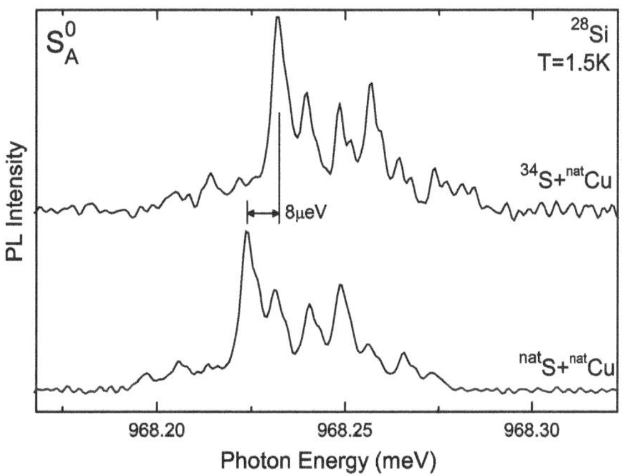

Fig. 4.20 The S_A centre in ^{28}Si for samples diffused with either ^{34}S or ^{nat}S (95 % ^{32}S) in combination with ^{nat}Cu, showing the S isotope shift

small amounts of luminescence are seen as a result of a small amount of ^{63}Cu in the sample, and were labeled "2", which indicates a combination of one ^{63}Cu and two ^{65}Cu. Above these spectra, a sample with a relatively high ^{65}Cu isotope content is shown and labeled ^{antinat}Cu in contrast to ^{nat}Cu, which consists of 69.3 % ^{63}Cu and is shown above. Disregarding the unfortunate broadening of the S_B centre in the "antinatural" sample, its spectrum is a mirror image of that for ^{nat}Cu. Comparing these four spectra, it was concluded that the complexes containing mixed Cu isotopes (labeled "1" and "2" in the figure) show a split pattern, as indicated by the braces in Fig. 4.21. Therefore, it appears that at least three Cu atoms must be assigned to the S_B Cu–S centre [2]. Using three as the number of Cu atoms, the Cu isotope shift for the S_B NP transition is ∼7.3 μeV/amu (based on the ^{63}Cu and ^{65}Cu spectra). The S_B centre could be viewed as an additional member of the four-atom deep PL centre family, but the inclusion of a chalcogen atom suggests that the details may be completely different.

Note that the PL components "0" and "3", and the individual components of "1" and "2", are split into identical triplets, with the central region of the spectrum

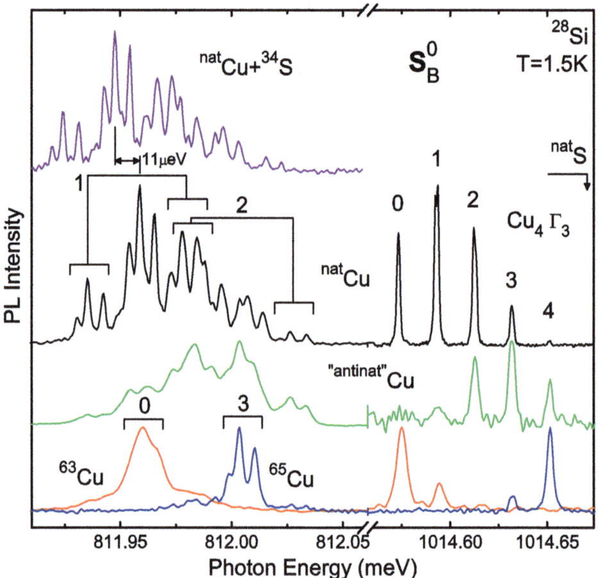

Fig. 4.21 The NP transitions of the S_B centre in ^{28}Si are shown on the *left*. The *top spectrum* is from a sample diffused with ^{34}S and ^{nat}Cu, whereas all the other spectra are from samples diffused with ^{nat}S (95 % ^{32}S). In the lower spectra, the S_B centre luminescence is shown, along with, on the *right*, that of the Γ$_3$ PL of the Cu$_4$ centre in the same samples, to clarify the Cu isotope content in the different samples. The *bottom* two spectra show the effect of using enriched single Cu isotopes, whereas the second one from the *bottom* has an isotopic composition that is skewed toward ^{65}Cu and, thus, labeled "antinatural". Labels 0–4 specify the number of ^{65}Cu atoms in a complex, where the S_B centre contains three Cu atoms. The three-component *brackets* indicate the three-fold electronic splitting of the S_B ground state [21], which is replicated in each isotopic component

being obscured by overlap. The energy splittings between the components of these repeated triplets have been measured to be 4.8 and 7.1 μeV [2], in agreement with the splittings measured for this centre using zero-field optically detected magnetic resonance (ODMR) [21].

References

1. M. Thewalt, M. Steger, A. Yang, M. Cardona, H. Riemann, N. Abrosimov, M. Churbanov, A. Gusev, A. Bulanov, I. Kovalev, A. Kaliteevskii, O. Godisov, P. Becker, H.-J. Pohl, Can highly enriched ^{28}Si reveal new things about old defects? Phys. B **401–402**, 587–592 (2007)
2. A. Yang, M. Steger, M. Thewalt, M. Cardona, H. Riemann, N. Abrosimov, M. Churbanov, A. Gusev, A. Bulanov, I. Kovalev, A. Kaliteevskii, O. Godisov, P. Becker, H.-J. Pohl, J. Ager III, E. Haller, High resolution photoluminescence of sulphur- and copper-related isoelectronic bound excitons in highly enriched ^{28}Si. Phys. B **401–402**, 593–596 (2007)
3. M. Steger, A. Yang, N. Stavrias, M. Thewalt, H. Riemann, N. Abrosimov, M. Churbanov, A. Gusev, A. Bulanov, I. Kovalev, A. Kaliteevskii, O. Godisov, P. Becker, H.-J. Pohl, Reduction of the Linewidths of Deep Luminescence Centers in ^{28}Si Reveals Fingerprints of the Isotope Constituents. Phys. Rev. Lett. **100**, 177402 (2008)
4. M. Steger, A. Yang, M.L.W. Thewalt, H. Riemann, N.V. Abrosimov, P. Becker, H.-J. Pohl, High resolution photoluminescence of copper, silver, gold and lithium-related isoelectronic bound excitons in highly enriched ^{28}Si, in *AIP Conference Proceeding*, ed. by M. Caldas, N. Studart, vol. 1199, pp. 33–34, 2010
5. M. Steger, A. Yang, T. Sekiguchi, K. Saeedi, M. Thewalt, M. Henry, K. Johnston, H. Riemann, N. Abrosimov, M. Churbanov, A. Gusev, A. Bulanov, I. Kaliteevski, O. Godisov, P. Becker, H.-J. Pohl, Isotopic fingerprints of gold-containing luminescence centers in ^{28}Si. Phys. B **404**, 5050–5053 (2009)
6. M. Steger, A. Yang, T. Sekiguchi, K. Saeedi, M.L.W. Thewalt, M.O. Henry, K. Johnston, E. Alves, U. Wahl, H. Riemann, N.V. Abrosimov, M.F. Churbanov, A.V. Gusev, A.K. Kaliteevskii, O.N. Godisov, P. Becker, H.-J. Pohl, Isotopic fingerprints of Pt-containing luminescence centers in highly enriched ^{28}Si. Phys. Rev. B **81**, 235217 (2010)
7. M. Steger, A. Yang, T. Sekiguchi, K. Saeedi, M.L.W. Thewalt, M.O. Henry, K. Johnston, H. Riemann, N.V. Abrosimov, M.F. Churbanov, A.V. Gusev, A.K. Kaliteevskii, O.N. Godisov, P. Becker, H.-J. Pohl, Photoluminescence of deep defects involving transition metals in Si- new insights from highly enriched ^{28}Si. Appl. Phys. Rev. **110**, 081301 (2011)
8. G. Davies, M. do Carmo, Vibronic coupling in shallow excited states of optical centres in silicon. Inst. Phys. Conf. Ser. **95**, 125–130 (1989)
9. N.T. Son, M. Singh, J. Dalfors, B. Monemar, E. Janzén, Electronic structure of a photoluminescent center in silver-doped silicon. Phys. Rev. B **49**, 17428–17431 (1994)
10. G. Davies, T. Gregorkiewicz, M. Zafar Iqbal, M. Kleverman, E.C. Lightowlers, N.Q. Vinh, M. Zhu, Optical properties of a silver-related defect in silicon. Phys. Rev. B **67**, 235111 (2003)
11. K. McGuigan, M. Henry, E. Lightowlers, A. Steele, M. Thewalt, A new photoluminescence band in silicon lightly doped with copper. Solid State Commun. **68**, 7–11 (1988)
12. J. Weber, H. Bauch, R. Sauer, Optical properties of copper in silicon: excitons bound to isoelectronic copper pairs. Phys. Rev. B **25**, 7688–7699 (1982)
13. R. Sauer, J. Weber, Photoluminescence characterisation of deep defects in silicon. Phys. B+C, **116**, 195 (1983)
14. L. Canham, G. Davies, E. Lightowlers, G. Blackmore, Complex isotope splitting of the no-phonon lines associated with exciton decay at a four-lithium-atom isoelectronic centre in silicon. Phys. B, **117, 118**, 119–121 (1983)
15. S. Watkins, U. Ziemelis, M. Thewalt, Long lifetime photoluminescence from a deep centre in copper-doped silicon. Solid State Commun. **43**, 687–690 (1982)

16. M. Henry, S. Daly, C. Frehill, E. MeGlynn, C. McDonagh, A photoluminescence study of gold- and platinum-related defects in silicon using radioactive transformations, in *Physics of Semiconductors: 23rd International Conference on the Physics of Semiconductors-ICPS 1996*, ed. by M. Scheffler, R. Zimmermann, vol. 1996, pp. 2713–2716, 1996

17. M.O. Henry, E. Alves, J. Bollmann, A. Burchard, M. Deicher, M. Fanciulli, D. Forkel-Wirth, M.H. Knopf, S. Lindner, R. Magerle, R. McGlynn, K.G. McGuigan, J.C. Soares, A. Stotzler, G. Weyer, Radioactive isotope identifications of Au and Pt photoluminescence centres in silicon. Phys. Stat. Solidi B **210**, 853 (1998)

18. M. Henry, M. Deicher, R. Magerle, E. McGlynn, A. Stotzler, Photoluminescence analysis of semiconductors using radioactive isotopes. Hyperf. Int. **129**, 443–460 (2000)

19. J.P. Leitão, M.C. Carmo, M.O. Henry, E. McGlynn, Uniaxial stress study of the 1026 meV center in Si:Pt. Phys. Rev. B **63**, 235208 (2001)

20. M. Singh, E.C. Lightowlers, G. Davies, C. Jeynes, K.J. Reeson, Isoelectronic bound exciton photoluminescence from a metastable defect in sulphur-doped silicon. Mat. Sci. Eng. B **4**, 303–307 (1989)

21. A.M. Frens, M.T. Bennebroek, J. Schmidt, W.M. Chen, B. Monemar, Zero-field optical detection of magnetic resonance on a metastable sulfur-pair-related defect in silicon: evidence for a Cu constituent. Phys. Rev. B **46**, 12316–12322 (1992)

Chapter 5
Discussion and Conclusion

5.1 Discussion of Experimental Results

In Chap. 2, all of the existing literature regarding these TM-containing deep PL centres in Si has been reviewed; Chap. 4 provides all of the details made available by high-resolution spectroscopy in highly enriched ^{28}Si, for both the well known and the newly discovered centres. These details include the atomic composition of the binding centre as revealed by the isotopic fingerprint, and the energies and isotope shifts of the NP transitions, including splittings of the IBE ground state, which results in different electronic transitions. The recorded relative intensities of these transitions indicate whether transitions are (at least partially) forbidden or allowed. Also provided are the energies, FWHM, and, where available, the isotope shifts of low energy pLVM and high energy LVM replicas of the NP transitions.

In this chapter, an overview of the newly available information is provided. We are looking for trends, and any other observations that may contribute toward finding a detailed model for the structure, formation, and properties of these large families of PL centres. To summarize, this information suggests a simple model in which a substitutional TM core, which can be Cu, Ag, Pt, or Au, sequentially binds positively charged Cu_i, Ag, or Li, as proposed by Shirai et al. [1, 2] and Estreicher and Carvalho [3, 4] for the Cu_4 centre. Even the Li_4 V_{Si} centre may be thought of as part of this family, with the V_{Si} taking the place of the substitutional TM. (Similar complexes of a V_{Si} with interstitial TM would not be expected, because the interstitial TM would immediately occupy the V_{Si}.) To generate the substitutional TM core of these IBE centres, a high temperature diffusion is needed and the quenching to lower temperature provides a supersaturation of the interstitial species, which remain mobile even at the lower temperature. Formation of the centres would be aided by long-range electrostatic interactions between the positively charged interstitials and a negatively charged core, even though the stability of the centres is known [5] to be greater than that provided by electrostatic interactions alone. Indeed, a naive explanation of the stability of the prototypical Cu_4 complex would be obvious if Cu_s in Si was a triple acceptor, as was originally suggested [6], and which is the case [6–8] for

Cu in Ge. Cu_s in Si is known to have singly charged donor and acceptor levels [9, 10], and a doubly charged acceptor level [11] (note that this level has also been assigned to a single donor level of Cu_i [12]). Unfortunately no triply charged level of Cu_s has been found [11].

Among the trends we can consider are the relative intensities, and presumably concentrations, of different centres, and the fact that not all possible members of the different families are observed. Before drawing any strong conclusions from this, it is important to realize that none of these PL centres is very stable and that they all anneal out at temperatures only moderately above room temperature. Indeed, the slow replacement of Cu with Li over a time scale of days to months indicates that even centres that are thought to be stable at room temperature may not in fact be truly stable, but rather losing and recapturing their interstitial components. Therefore the failure to observe certain complexes may not indicate substantial differences in properties, but instead small differences in relative stabilities.

Given the above substitutional-interstitial model, the families of impurity complexes can be grouped by their substitutional core impurity. For Cu_{Si}, only the four-atom Cu_4 centre has ever been observed by PL and there is no evidence of PL from centres with interstitial Li substituting for Cu.

For centres with an Ag_{Si} core, again only four-atom complexes have been observed. The Ag_4 centre is only strong in samples quenched in a sealed ampoule after diffusion with excess Ag present, which perhaps acts as a getter, thereby reducing the concentration of other TM in the sample. The interstitial Ag is rapidly replaced by Cu after subsequent quenching, even if efforts are made to reduce Cu contamination. This results in strong Cu_3Ag PL, and much weaker Cu_2Ag_2 PL, whereas PL from a $CuAg_3$ complex has never been observed. In the presence of Li, strong PL from the $CuLi_2Ag$ centre is observed, but there is no evidence for any other complexes that contain Ag, Cu, and Li.

For centres with an Au_{Si} core, both four-atom and five-atom complexes are observed, with PL from the four-atom complexes typically being much stronger, and with a larger binding energy (lower PL energy). With the exception of the missing Li_4Au centre, all possible substitutions of Li for Cu are observed, and the sequential substitution of Li for Cu is seen to increase the energy of the NP transition, as summarized in Fig. 5.1.

The behaviours described above are reversed for centres that have a Pt_{Si} core with the PL of five-atom complexes being lower in energy and much stronger than that of the four-atom centres, and with substitutions of Li for Cu now decreasing the NP PL energy, as seen in Fig. 5.1.

It should be emphasized at this point that although the properties of the centres reviewed in Chap. 2 had strong similarities—namely donor-like IBE character and trigonal symmetry—most of the newly discovered centres have not been studied using uniaxial stress or Zeeman spectroscopy, nor has their excited state structure been investigated. In the absence of further experimental studies or detailed modelling of the properties of the binding centres, it would be premature to assume that all of the new centres also have trigonal symmetry and donor-like IBE character.

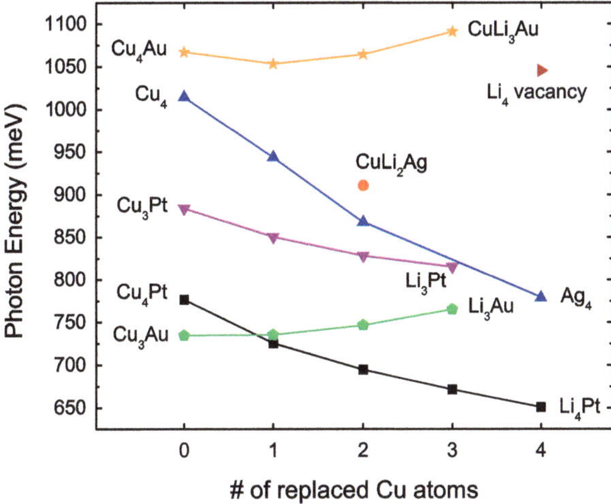

Fig. 5.1 In this diagram the energies of the no-phonon transitions of different centres are shown versus the number of Cu atoms in the complex that have been replaced with Li or Ag. Lines connect centres of the same series. For example, it can be seen, that the 4 and 5-atom Au related centres follow a similar trend, as do the Pt-related centres, albeit in opposite directions with Li substitutions. For reference, the CuLi$_2$Ag and Li$_4$ V$_{Si}$ [13] centres are also shown

The new model of Estreicher et al. [3] and Carvalho et al. [4] for the Cu$_4$ centre which we have introduced in Sect. 2.1 may in future provide a path to understanding the dominance of four- and five-atom centres with a substitutional core other than Cu. In this context it is again important to emphasize that the material used here, and to the best of our knowledge in previous studies, was lightly doped.

5.2   Conclusion

This thesis gives an overview of recent discoveries on isoelectronically bound exciton centres in silicon. First, studies of the properties of many of the centres discussed here, some going back more than 30 years, have been reviewed. After discussing experimental methods and the sample preparation, the novel characterization method of isotopic fingerprints in photoluminescence spectroscopy in ^{28}Si was introduced. The new results include impurity complexes of the constituents Li, S, Cu, Ag, Pt, and Au. Surprisingly, it turns out that none of the previously studied complexes are what they were thought to be. The differences range from a mere deviation in the number of involved impurity atoms, to there being no overlap at all between the previously assumed constituents and those revealed in ^{28}Si.

In addition, the ^{28}Si studies revealed many new TM-containing PL defects, which seem to fall into families of four-atom and five-atom complexes. We suggest these complexes have a substitutional core of a single Cu, Ag, Pt, or Au atom surrounded by three or four interstitial impurities, which can be Cu, Ag, or Li.

For most of these newly discovered centres, the fundamental properties, such as binding centre symmetry and donor-like versus acceptor-like IBE properties, are not yet known. It should not be assumed that these centres are of the same symmetry as the known centres, so there is still room for further experimental studies. What is needed most at this time, however, is some theoretical understanding of the properties and formation mechanisms of these ubiquitous PL defects, especially for the previously known centres in light of the recent discoveries. It is encouraging that work has already begun [1–4, 14–16] on the prototypical Cu_4 centre, but a more detailed understanding of the stabilities, IBE properties, and detailed formation mechanism is still needed. It is also unclear why the centres with Cu and Ag cores have only four-atom members, but the Au and Pt centres both have four-atom and five-atom families. The information gathered here, which will also be available in the form of a review publication [17], will hopefully encourage further theoretical studies and modelling on these interesting PL centres.

The isotopic fingerprint studies presented here have thus far been limited to those PL centres for which there was good reason to suspect the involvement of TM, since ^{28}Si is a scarce resource, and parts of the available material are dedicated to other research fields [18–21]. Nevertheless, it would be interesting to investigate the isotopic fingerprints of other IBE centres in ^{28}Si, such as those suggested to involve Al and N [22–24], Mn [25], Zn [25, 26], Pd [27], Cd [28], In [29–33], Hg [34, 35], and Tl [31, 36, 37].

References

1. K. Shirai, H. Yamaguchi, A. Yanase, H. Katayama-Yoshida, A new structure of Cu complex in Si and its photoluminescence. J. Phys. Cond. Mat. **21**, 064249 (2009)
2. K. Shirai, H. Yamaguchi, J. Ishisada, K. Matsukawa, A. Yanase, S. Emura, Cu complex in silicon and its photoluminescence, in *AIP Conference Proceedings*, vol. 1199, ed. by M. Caldas, N. Studart (American Institute of Physics, Melville, 2010), pp. 91–92
3. S.K. Estreicher, A. Carvalho, The Cu_{PL} defect and the $Cu_{s1}Cu_{i3}$ complex. Phys. B: Cond. Mat. **407**(15), 2967–2969 (2012)
4. A. Carvalho, D.J. Backlund, S.K. Estreicher, Four-copper complexes in Si and the Cu_{PL} defect: a first-principles study. Phys. Rev. B **84**, 155322 (2011)
5. A.A. Istratov, H. Hieslmair, T. Heiser, C. Flink, E.R. Weber, The dissociation energy and the charge state of a copper-pair center in silicon. Appl. Phys. Lett. **72**, 474–476 (1998)
6. R.N. Hall, J.H. Racette, Diffusion and solubility of copper in extrinsic and intrinsic germanium, silicon, and gallium arsenide. J. Appl. Phys. **35**, 379–397 (1964)
7. H.H. Woodbury, W.W. Tyler, Triple acceptors in germanium. Phys. Rev. **105**, 84–92 (1957)
8. M.K. Nissen, A.G. Steele, M.L.W. Thewalt, Photoluminescence from excitons bound to a triple acceptor, Ge:Cu. Phys. Rev. B **41**, 7926–7928 (1990)
9. S. Knack, J. Weber, H. Lemke, H. Riemann, Evolution of copper-hydrogen-related defects in silicon. Phys. B **308–310**, 404–407 (2001)
10. N. Yarykin, J. Weber, Copper-related deep-level centers in irradiated p-type silicon. Phys. Rev. B **83**, 125207 (2011)

11. S. Knack, Copper-related defects in silicon. Mat. Sci. Semicond. Process. **7**, 125–141 (2004)
12. A.A. Istratov, H. Hieslmair, C. Flink, T. Heiser, E.R. Weber, Interstitial copper-related center in n-type silicon. Appl. Phys. Lett. **71**, 2349–2351 (1997)
13. G. Davies, L. Canham, E. Lightowlers, Magnetic and uniaxial stress perturbations of optical transitions at a four Li atom complex in Si. J. Phys. C **17**, L173–L178 (1984)
14. M. Nakamura, S. Murakami, N.J. Kawai, S. Saito, K. Matsukawa, H. Arie, Compositional Transformation between Cu centers by annealing in Cu-diffused silicon crystals studied with deep-level transient spectroscopy and photoluminescence. Jpn. J. Appl. Phys. **48**, 082302 (2009)
15. M. Nakamura, S. Murakami, Deep-level transient spectroscopy and photoluminescence studies of formation and depth profiles of copper centers in silicon crystals diffused with dilute copper. Jpn. J. Appl. Phys. **49**, 071302 (2010)
16. M. Nakamura, S. Murakami, Depth progression of dissociation reaction of the 1.014-eV photoluminescence copper center in copper-diffused silicon crystal measured by deep-level transient spectroscopy. Appl. Phys. Lett. **98**, 141909 (2011)
17. M. Steger, A. Yang, T. Sekiguchi, K. Saeedi, M.L.W. Thewalt, M.O. Henry, K. Johnston, H. Riemann, N.V. Abrosimov, M.F. Churbanov, A.V. Gusev, A.K. Kaliteevskii, O.N. Godisov, P. Becker, H.-J. Pohl, Photoluminescence of deep defects involving transition metals in Si—new insights from highly enriched ^{28}Si. Appl. Phys. Rev. **110**, 081301 (2011)
18. A. Yang, M. Steger, M.L.W. Thewalt, T.D. Ladd, K.M. Itoh, E.E. Haller, J.W. Ager III, H. Riemann, N.V. Abrosimov, P. Becker, H.-J. Pohl, Nuclear polarization of phosphorus donors in ^{28}Si by selective optical pumping, in *AIP Conference Proceeding*, vol. 1199, ed. by M. Caldas, N. Studart (American Institute of Physics, Melville, 2010), pp. 375–376
19. M. Steger, T. Sekiguchi, A. Yang, K. Saeedi, M.E. Hayden, M.L.W. Thewalt, K.M. Itoh, H. Riemann, N.V. Abrosimov, P. Becker, H.-J. Pohl, Optically detected NMR of optically hyperpolarized ^{31}P neutral donors in ^{28}Si. J. Appl. Phys. **109**, 102411 (2011)
20. S. Simmons, R.M. Brown, H. Riemann, N.V. Abrosimov, P. Becker, H.-J. Pohl, M.L.W. Thewalt, K.M. Itoh, J.J.L. Morton, Entanglement in a solid-state spin ensemble. Nature **470**, 69–72 (2011)
21. M. Steger, K. Saeedi, M.L.W. Thewalt, J.J.L. Morton, H. Riemann, N.V. Abrosimov, P. Becker, H.-J. Pohl, Quantum information storage for over 180 s using donor spins in a ^{28}Si "semiconductor vacuum". Science **336**, 1280–1283 (2012)
22. R.A. Modavis, D.G. Hall, Aluminum-nitrogen isoelectronic trap in silicon. J. Appl. Phys. **67**, 545–547 (1990)
23. J. Weber, W. Schmid, R. Sauer, Excitons bound to an isoelectronic trap in silicon. J. Lumin. **18–19**, 93–96 (1979)
24. R. Sauer, J. Weber, W. Zulehner, Nitrogen in silicon: towards the identification of the 1.1223-eV (A, B, C) photoluminescence lines. Appl. Phys. Lett. **44**, 440–442 (1984)
25. M.O. Henry, K.G. McGuigan, R.C. Barklie, Bound exciton recombination at Mn-Zn pair centres in silicon. Solid State Commun. **64**, 31–33 (1987)
26. M.O. Henry, D.J. Beckett, A.G. Steele, M.L.W. Thewalt, K.G. McGuigan, A zinc-related isoelectronic bound exciton in silicon. Solid State Commun. **66**, 689–694 (1988)
27. V. Thomas, J. Barrau, M. Brousseau, K. Kirouni, G. Couloumiers, The luminescence of palladium-doped silicon. Phys. Stat. Sol. B **148**, 723–727 (1988)
28. E. McGlynn, M.O. Henry, K.G. McGuigan, M.C. doCarmo, Photoluminescence study of cadmium-related defects in oxygen-rich silicon. Phys. Rev. B **54**, 14494–14503 (1996)
29. G.S. Mitchard, S.A. Lyon, K.R. Elliott, T.C. McGill, Observation of long lifetime lines in photoluminescence from Si:In. Solid State Commun. **29**, 425–429 (1979)
30. M.L.W. Thewalt, U.O. Ziemelis, P.R. Parsons, Enhancement of long lifetime lines in photoluminescence from Si:In. Solid State Commun. **39**, 27–30 (1981)
31. T.E. Schlesinger, T.C. McGill, Role of Fe in new luminescence lines in Si:Tl and Si:In. Phys. Rev. B **25**, 7850–7851 (1982)
32. T. Schlesinger, R. Hauenstein, R. Feenstra, T. McGill, Isotope shifts for the P, Q, R lines in indium-doped silicon. Solid State Commun. **46**, 321–324 (1983)

33. S. Watkins, M. Thewalt, T. Steiner, Isoelectronic bound excitons in In- and Tl-doped Si: a novel semiconductor defect. Phys. Rev. B **29**, 5727–5738 (1984)
34. A. Henry, B. Monemar, J.P. Bergman, J.L. Lindström, P.O. Holtz, Y. Zhang, J.W. Corbett, Mercury-related luminescent center in silicon. Phys. Rev. B **47**, 13309–13313 (1993)
35. M.O. Henry, E. McGlynn, J. Fryar, S. Lindner, J. Bollmann, The evolution of point defects in semiconductors studied using the decay of implanted radioactive isotopes. Nucl. Instr. Meth. Phys. Res. B **178**, 256–259 (2001)
36. M.L.W. Thewalt, U.O. Ziemelis, R.R. Parsons, Isoelectronic bound excitons in silicon: the role of deep acceptors. Phys. Rev. B **24**, 3655–3658 (1981)
37. H. Conzelmann, A. Hangleiter, J. Weber, Thallium-related isoelectronic bound excitons in silicon. A bistable defect at low temperatures. Phys. Stat. Sol. B **133**, 655–668 (1986)

Curriculum Vitae

Michael Steger started his career in Physics at the University of Augsburg, Germany in 2001. He joined the graduate physics program at Simon Fraser University (Vancouver, Canada) in 2004 and obtained his MSc in Physics in 2007. Michael continued his PhD studies at Simon Fraser University and defended his PhD thesis in 2011. As a recognition for his outstanding PhD research and academic excellence Michael was awarded the 'Dean of Graduate Studies Convocation Medal' by Simon Fraser University's senate. During his graduate studies Michael has published extensively in peer-reviewed journals. He received two "Editor's Sugestion" awards and the invitation to publish an Applied Journal of Physics Review article. Michael also presented at numerous international conferences.

List of Publications

1 M. Steger, K. Saeedi, M. L. W. Thewalt, J. J. L. Morton, H. Riemann, N. V. Abrosimov, P. Becker, and H.-J. Pohl, "Quantum Information Storage for over 180 s Using Donor Spins in a ^{28}Si "Semiconductor Vacuum"", Science **336**, 1280–1283 (2012).

M. Steger, *Transition-Metal Defects in Silicon*, Springer Theses, DOI: 10.1007/978-3-642-35079-5, © Springer-Verlag Berlin Heidelberg 2013

2 D. Lackner, M. Steger, M. L. W. Thewalt, O. J. Pitts, Y. T. Cherng, S. P. Watkins, E. Plis, and S. Krishna, "InAs/inAsSb strain balanced superlattices for optical detectors: Material properties and energy band simulations", J. Appl. Phys. **111**, 034507, 034507 (2012).

3 M. Steger, T. Sekiguchi, A. Yang, K. Saeedi, M. E. Hayden, M. L. W. Thewalt, K. M. Itoh, H. Riemann, N. V. Abrosimov, P. Becker, and H.-J. Pohl, "Optically detected NMR of optically hyperpolarized ^{31}P neutral donors in ^{28}Si", J. Appl. Phys. **109**, 102411, 102411, (2011).

4 M. Steger, A. Yang, T. Sekiguchi, K. Saeedi, M. L. W. Thewalt, M. O. Henry, K. Johnston, H. Riemann, N. V. Abrosimov, M. F. Churbanov, A. V. Gusev, A. K. Kaliteevskii, O. N. Godisov, P. Becker, and H.-J. Pohl, "Photoluminescence of deep defects involving transition metals in Si – new insights from highly enriched ^{28}Si", App. Phys. Rev. **110**, 081301 (2011).

5 M. Steger, A. Yang, T. Sekiguchi, K. Saeedi, M. L. W. Thewalt, M. O. Henry, K. Johnston, E. Alves, U. Wahl, H. Riemann, N. V. Abrosimov, M. F. Churbanov, A. V. Gusev, A. K. Kaliteevskii, O. N. Godisov, P. Becker, and H.-J. Pohl, "Isotopic fingerprints of Pt-containing luminescence centers in highly enriched ^{28}Si", Phys. Rev. B **81**, 235217 (2010).

6 T. Sekiguchi, M. Steger, K. Saeedi, M. L. W. Thewalt, H. Riemann, N. V. Abrosimov, and N. Nötzel, "Hyperfine Structure and Nuclear Hyperpolarization Observed in the Bound Exciton Luminescence of Bi Donors in Natural Si". Phys. Rev. Lett. **104**, 137402 (2010).

7 D. Lackner, M. Martine, Y. T. Cherng, M. Steger, W. Walukiewicz, M. L. W. Thewalt, P. M. Mooney, and S. P. Watkins, "Electrical and optical characterization of n-InAsSb/n-GaSb heterojunctions", J. Appl. Phys. **107**, 014512, 014512 (2010).

8 M. Steger, A. Yang, M. L. W. Thewalt, H. Riemann, N. V. Abrosimov, P. Becker, and H.-J. Pohl, "High Resolution Photoluminescence of Copper, Silver, Gold and Lithium-related Isoelectronic Bound Excitons in Highly Enriched ^{28}Si", AIP Conf. Proc. **1199**, edited by M. Caldas and N. Stedart, 33–34, (2010).

9 A. Yang, M. Steger, M. L. W. Thewalt, T. D. Ladd, K. M. Itoh, E. E. Haller, J. W. Ager III, H. Riemann, N. V. Abrosimov, P. Becker, and H.-J. Pohl, "Nuclear Polarization of Phosphorus Donors in ^{28}Siby Selective Optical Pumping", AIP Conf. Proc., **1199** edited by M. Caldas and N. Stedart, 375–376, (2010).

10 M. Steger, A. Yang, M. L. W. Thewalt, M. Cardona, H. Riemann, N. V. Abrosimov, M. F. Churbanov, A. V. Gusev, A. D. Bulanov, I. D. Kovalev, A. K. Kaliteevskii, O. N. Godisov, P. Becker, H.-J. Pohl, E. E. Haller, and J. W. Ager III, "High-resolution absorption spectroscopy of the deep impurities S and Se in ^{28}Si revealing the ^{77}Se hyperfine splitting", Phys. Rev. B, **80**, 115204, 115204 (2009).

11 M. Steger, A. Yang, T. Sekiguchi, K. Saeedi, M. L. W. Thewalt, M. Henry, K. Johnston, H. Riemann, N. V. Abrosimov, M. F. Churbanov, A. V. Gusev, A. D. Bulanov, I. D. Kaliteevski, O. N. Godisov, P. Becker, and H.-J. Pohl, "Isotopic fingerprints of gold-containing luminescence centers in ^{28}Si", Physica B **404**, 5050–5053, 2009.

12 M. Steger, A. Yang, D. Karaiskaj, M. L. W. Thewalt, E. E. Haller, J. W. Ager III, M. Cardona, H. Riemann, N. V. Abrosimov, A. V. Gusev, A. D. Bulanov, A. K. Kaliteevskii, O. N. Godisov, P. Becker, and H.-J. Pohl, "Shallow impurity absorption spectroscopy in isotopically enriched silicon", Phys. Rev. B **79** 205210, 205210 (2009).

13 A. Yang, M. Steger, T. Sekiguchi, M. L. W. Thewalt, J. W. Ager III, and E. E. Haller, "Homogeneous linewidth of the 31P bound exciton transition in silicon", Appl. Phys. Lett. **95** 122113, 2009.

14 A. Yang, M. Steger, T. Sekiguchi, M. L. W. Thewalt, T. D. Ladd, K. M. Itoh, H. Riemann, N. V. Abrosimov, P. Becker, and H.-J. Pohl. "Simultaneous Subsecond Hyperpolarization of the Nuclear and Electron Spins of Phosphorus in Silicon by Optical Pumping of Exciton Transitions", Phys. Rev. Lett. **102**, 257401, 257401 (2009).

15 A. Yang, M. Steger, T. Sekiguchi, D. Karaiskaj, M. L. W. Thewalt, M. Cardona, K. M. Itoh, H. Riemann, N. V. Abrosimov, M. F. Churbanov, A. V. Gusev, A. D. Bulanov, I. D. Kovalev, A. K. Kaliteevskii, O. N. Godisov, P. Becker, H.-J. Pohl, J. W. Ager, and E. E. Haller, "Single-frequency laser spectroscopy of the boron bound exciton in ^{28}Si" Phys. Rev. B **80** 195203 (2009).

16 D. Lackner, O. J. Pitts, M. Steger, A. Yang, M. L. W. Thewalt, and S. P. Watkins, "Strain balanced inas/inassb superlattice structures with optical emission to 10 μm", Appl. Phys. Lett. **95**, 081906, 081906 (2009).

17 D. Lackner, O. J. Pitts, S. Najmi, P. Sandhu, K. L. Kavanagh, A. Yang, M. Steger, M. L. W. Thewalt, Y. Wang, D. W. McComb, C. R. Bolognesi, and S. P. Watkins, "Growth of InAsSb/InAs MQWs on GaSb for mid-IR photodetector applications", J. Cryst. Growth, **311**, 3563–3567 (2009).

18 M. Steger, A. Yang, N. Stavrias, M. L. W. Thewalt, H. Riemann, N. V. Abrosimov, M. F. Churbanov, A. V. Gusev, A. D. Bulanov, I. D. Kovalev, A. K. Kaliteevskii, O. N. Godisov, P. Becker, and H.-J. Pohl, "Reduction of the Linewidths of Deep Luminescence Centers in ^{28}Si Reveals Fingerprints of the Isotope Constituents", Phys. Rev. Lett. **100**, 177402, 177402 (2008).

19 A. Yang, M. Steger, H. J. Lian, M. L. W. Thewalt, M. Uemura, A. Sagara, K. M. Itoh, E. E. Haller, J. W. Ager, S. A. Lyon, M. Konuma, and M. Cardona, "High-resolution photoluminescence measurement of the isotopic-mass dependence of the lattice parameter of silicon", Phys. Rev. B **77**, 113203 (2008).

20 M. Steger, A. Yang, M. L. W. Thewalt, M. Cardona, H. Riemann, N. V. Abrosimov, M. F. Churbanov, A. V. Gusev, A. D. Bulanov, I. D. Kovalev, A. K. Kaliteevskii, O. N. Godisov, P. Becker, H.-J. Pohl, J. W. Ager III, and

E. E. Haller, "Impurity Absorption Spectroscopy of the Deep Double Donor Sulfur in Isotopically Enriched Silicon", Physica B **401–402**, 600–603 (2007).

21 M. L. W. Thewalt, M. Steger, A. Yang, M. Cardona, H. Riemann, N. V. Abrosimov, M. F. Churbanov, A. V. Gusev, A. D. Bulanov, I. D. Kovalev, A. K. Kaliteevskii, O. N. Godisov, P. Becker, and H.-J. Pohl, "Can highly enriched ^{28}Si reveal new things about old defects?" Physica B, **401–402**, 587–592 (2007).

22 M. L. W. Thewalt, A. Yang, M. Steger, D. Karaiskaj, M. Cardona, H. Riemann, N. V. Abrosimov, A. V. Gusev, A. D. Bulanov, I. D. Kovalev, A. K. Kaliteevskii, O. N. Godisov, P. Becker, H. J. Pohl, E. E. Haller, J. W. Ager III, and K. M. Itoh, "Direct observation of the donor nuclear spin in a near-gap bound exciton transition: ^{31}P in highly enriched ^{28}Si", J. Appl. Phys. **101**, 081724, 081724 (2007).

23 A. Yang, M. Steger, M. L. W. Thewalt, M. Cardona, H. Riemann, N. V. Abrosimov, M. F. Churbanov, A. V. Gusev, A. D. Bulanov, I. D. Kovalev, A. K. Kaliteevskii, O. N. Godisov, P. Becker, H.-J. Pohl, J. W. Ager III, and E. E. Haller, "High resolution photoluminescence of sulphur- and copper-related isoelectronic bound excitons in highly enriched ^{28}Si", Physica B, **401–402**, 593–596 (2007).

24 M. Steger, A. Yang, D. Karaiskaj, M. L. W. Thewalt, E. E. Haller, J. W. Ager III, M. Cardona, H. Riemann, N. V. Abrosimov, A. V. Gusev, A. D. Bulanov, A. K. Kaliteevskii, O. N. Godisov, P. Becker, H.-J. Pohl, and K. M. Itoh, "Shallow Impurity Absorption Spectroscopy in Isotopically Enriched Silicon", AIP Conf. Proc. **893**, edited by W. Jantsch and F. Schaffler, 231–232 (2007).

25 A. Yang, M. Steger, D. Karaiskaj, M. L. W. Thewalt, M. Cardona, K. M. Itoh, H. Riemann, N. V. Abrosimov, M. F. Churbanov, A. V. Gusev, A. D. Bulanov, A. K. Kaliteevskii, O. N. Godisov, P. Becker, H.-J. Pohl, J. W. Ager III, and E. E. Haller, "Optical Detection and Ionization of Donors in Specific Electronic and Nuclear Spin States", Phys. Rev. Lett., **97**, 227401, 227401 (2006).

26 S. Najmi, X. K. Chen, A. Yang, M. Steger, M. L. W. Thewalt, and S. P. Watkins, "Local vibrational mode study of carbon-doped *InAs*". Phys. Rev. B **74**, 113202 (2006).

Conference Presentations Without Proceedings

- "Optically detected NMR of optically hyperpolarized ^{31}P neutral donors in ^{28}Si", International Workshop on Silicon Quantum Electronics 2011, Denver, CO

- "Isotopic Fingerprints of 4 and 5-atom Pt containing PL centers in ^{28}Si", Gordon Research Conference on Defects in Semiconductors 2010, New London, NH
- "Isotopic Fingerprints of four and five-atom Pt luminescence centers in ^{28}Si", APS March Meeting 2010, Portland, OR